BIOCHEMISTRY RESEARCH TRENDS

INTERPOLYELECTROLYTE COMPLEXES OF CHITOSAN

BIOCHEMISTRY RESEARCH TRENDS

Additional books in this series can be found on Nova's website
under the Series tab.

Additional E-books in this series can be found on Nova's website
under the E-book tab.

BIOCHEMISTRY RESEARCH TRENDS

INTERPOLYELECTROLYTE COMPLEXES OF CHITOSAN

N.A. SAMOILOVA
M.A. KRAYUKHINA
AND
I.A. YAMSKOV

NOVINKA

Nova Science Publishers, Inc.
New York

For permission to use material from this book please contact us:
Telephone 631-231-7269; Fax 631-231-8175
Web Site: http://www.novapublishers.com

LIBRARY OF CONGRESS CATALOGING-IN-PUBLICATION DATA

Samoilova, N. A. (Nadezhda A.), 1945-
Interpolyelectrolyte complexes of chitosan / N.A. Samoilova, M.A. Krayukhina, I.A. Yamskov.
p. ; cm.
Includes bibliographical references and index.
ISBN 978-1-61761-194-0 (softcover)
1. Chitosan--Biotechnology. 2. Chitosan. I. Krayukhina, M. A. (Maria A.), 1976- II. Yamskov, I. A. (Igor A.), 1944- III. Title.
[DNLM: 1. Chitosan--chemistry. 2. Electrolytes--chemistry. QU 83]
TP248.65.C55S26 2010
660.6--dc22

2010029767

Published by Nova Science Publishers, Inc. † New York

CONTENTS

PREFACE

The results of studies of interpolyelectrolyte complexes of chitosan and different polyanions are summarized and described systematically. Specific properties of chitosan as polyelectrolyte are described. The general concept of the formation of polyelectrolyte complexes is developed. Separate parts of the review are dedicated to investigation of polyelectrolyte complexes of chitosan with:

- biopolyelectrolytes, including polysaccarides (plant, animal, bacterial polysaccarides and lipopolysaccarides), proteins, nucleic acids and also modified natural polyanions (carboxymethylchitin, carboxymetylcellulose, etc.),
- synthetic polyanions.

The data on application of polyelectrolyte complexes of chitosan in medicine and biotechnology, particularly for creation of hemocompatible, thromboresistant materials, bioconstruction materials for replacement of coverlets, blood vessels, bone tissue, immobilization of biological active compounds, non-viral vectors of genetic information transfer and so on are generalized.

The bibliography includes 215 references.

Chapter 1

INTRODUCTION

In recent years, several publications reviewed data on complexes of the polysaccharide chitosan (CS) with anionic polyelectrolytes (mainly of natural origin) and considered the potential applications of chitosan compositions in bio-medicine, pharmacy, food industry, etc. [1-4]. In the present chapter is summarized literature data on the synthesis, study and application of polyelectrolyte complexes of chitosan with natural, natural modified and synthetic polyanions.

Chitosan has unique physicochemical and biological properties such as biocompatibility, antimicrobial activit, the ability to activate macrophages and enhance fibroblast proliferation, etc. Due to the diverse, recyclable and unlimited sources of the raw material, this biopolymer offers some important benefits [5, 6].

One of the attractive characteristics of chitosan is its ability to form non-covalent complexes with other polyelectrolytes. These complexes possess potentially useful specific properties [2]. Complexation is greatly affected by molecular mass of chitosan, degree of polydispersity, degree of deacetylation (DD), dissociation constant (pK_a) of the protonated amino groups of chitosan and the distribution of functional (amino and acetamido) groups along the polymer chain. The degree of deacetylation of chitosan determines the value of pK_a, which can vary from 6.46 to 7.32 [7]. The conformation state and size of chitosan macromolecules in solution also depend on the DD and solution ionic strength.

For DD > 80%, the Coulomb repulsion of the protonated amino groups of chitosan dominates within a certain range of pH and ionic strength. For 50% < DD < 80%, hydrogen bonds formed with participation of acetamide groups of chitosan make an additional contribution to intermolecular interactions. The bulky acetamide residues and the hydrogen bond network create steric hindrance and decelerate the rotation of the pyranose units around the β-glycosidic bond, increasing the rigidity of the polysaccharide macromolecule. As DD decreases, these effects become more important, and hydrophobic interaction increase; all this leads to increased self-aggregation of chitosan molecules in solutions, which also depend on the polymer concentration [4]. Thus, the ability of chitosan to form polyelectrolyte complexes depends on the nature of interactive polyelectrolytes and reaction conditions.

Chapter 2

GENERAL CONCEPT OF THE FORMATION OF POLYELECTROLYTE COMPLEXES

The complex systems formed by macromolecules (interpolymer complexes) can be classified according to the dominant type of intermolecular interaction:

- complexes formed by the van der Waals interactions,
- complexes formed by hydrogen bonding between macro molecules,
- coordination complexes,
- covalent complexes,
- polyelectrolyte complexes stabilized by intermolecular ionic bonds.
- sterically matching ('key – lock' type) complexes.

In addition, there are combined types of interpolymer complexes, which owe their stability to cooperative interactions involving the above-mentioned types of bonding. The overwhelming majority of publications are devoted to polyelectrolyte interactions at the level of linear, dendritic, protein, etc. macromolecules. Moreover, complexation of polyelectrolytes with oppositely charged surfaces (e.g., with polymer 'brushes' [8], latex particles and monolamellar liposomes [9, 10]) and macroscopic, weakly crosslinked hydrogels [11, 12] was described.

Several recent publications dealt with the properties of systems formed by the interaction of oppositely charged polyelectrolyte networks (microgel particles) [13] and nano-sized particles [14].

The ionogenic groups of the macromolecular chain of polyelectrolytes dissociate in polar media into charged units on the polymer chain and low-molecular-mass counterions. The electrostatic attraction of oppositely charged polymer units leads to the formation of polyion pairs, namely, interpolyelectrolyte complexes. In this review for simplicity we will use the therm polyelectrolyte complex and abbreviation PEC. The foundations of the theory of PEC formation and organization were laid in works [15 - 19].

Polyelectrolyte complexes have attracted the attention of researchers because of their unique properties and ease of preparation. PEC solutions can be regarded as smart polymer systems due to their ability to change their phase state after insignificant changes in the external factors (pH, ionic strength, temperature, etc.) [1, 20]. Some of the possible approaches to the classification of PEC are as follows:

- according to the nature of polyions involved in the formation of PEC, one can distinguish complexes based on polymers of only natural or synthetic origin and PEC of mixed type, involving synthetic polymers and biomacromo-lecules;
- complexes can be formed by strong or weak polyelectrolytes;
- according to their composition, PEC may be stoichiometric and non-stoichiometric;
- soluble and insoluble PEC can be formed depending on the reaction conditions.

A distinction of the interactions of polyelectrolytes (compared with processes involving their low molecular mass analogues) is the cooperative character of bonds formed between the electrostatically complementary anionogenic and cationogenic macromolecules involved in complexation. This type of bond imparts high stability to PEC within wide ranges of pH and ionic strength of the medium.

The formation of PEC and variation of its composition (i.e., the polyelectrolyte reaction) is a reversible process controlled by many factors such as the nature of polyelectrolytes, their molecular mass and molecular mass distribution, ionic strength, pH of polyelectrolyte solutions, etc. The same factors and the sequence in which the components are mixed regulate the formation of soluble and insoluble as well as stoichiometric and non-stoichiometric PEC. There are several procedures for the preparation of PEC. The simplest method is mixing of the aqueous solutions of polycations and polyanions. Another way is polymerisation of ionic monomers on oppositely charged polyionic matrices [15].

Upon mixing of solutions containing oppositely charged polyions, PECs are formed by an ionexchange reaction, occurring at a high rate even in highly dilute starting solution.

If complexation involves weak polyelectrolytes (for example, polyacrylic acid and poly(dimethylaminoethyl methacrylate)), the extent of reaction (conversion of ionogenic groups to the so-called interchain 'salt' bonds between two polyions) can be varied by changing the pH of the reaction system. In this case, the measure of the free energy of complexation (ΔG_{cs}) is ΔpH, which is the function of ionicity (α) and conversion (θ) of polyelectrolytes that defines the multicentre cooperative interaction between the oppositely charged polyion units for $\theta = \alpha$ (i.e., if each ionogenic group of co-polyelectrolytes of PEC is assumed to form, after being charged, an ionic bond with the corresponding oppositely charged group) [15]:

$$\Delta pH(\alpha,\theta) = [\Delta G(\alpha) - \Delta G(\theta)]/2.3RT = \Delta G_{cs}/2.3RT,$$

where $\Delta G(\alpha)$ and $\Delta G(\theta)$ are the free energies of ionization of a weak polyelectrolyte in the absence and the presence of an oppositely charged polyion for the given α and θ.

If a PEC is formed upon matrix polymerization of charged monomers on the oppositely charged polyion (matrix), the change in the free energy (motive force of complexation) is described by the equation:

$$\Delta G_{tp} = \Delta G_p + \Delta G_{cs},$$

where ΔG_p – is the free energy of polymerization in the absence of a matrix. When monomer polymerization is thermodynamically forbidden ($\Delta G_p > 0$), but ΔG_{cs} is negative and и $|\Delta G_{cs}| > |\Delta G_p|$, the matrix polyion shifts the equilibrium toward the oppositely charged polyion.

Note that the same factor is responsible for clusterization of protein globules with oppositely charged polyions to form protein polyelectrolyte complexes. When the ratio of the numbers of ionized groups (Z = m: n) of the starting polyelectrolytes is 1 and $\theta \rightarrow 1$, stoichiometric PEC are formed in the system, which are insoluble in any solvents, but can swell in water to a certain extent. These insoluble PECs resemble cross-linked hydrogels in their properties. They have found wide use as binding agents to prevent wind and water erosion of soil in agriculture; as effective coagulants of colloidal dispersions in in agriculture; as effective coagulants of colloidal dispersions in industry; as sorbents and matrices for enzyme immobilization and bio/hemocompatible coatings for medical materials contacting with blood and other biological media in medicine and biotechnology [1, 2, 5, 15].

Despite the long history of PEC studies, the structure and properties of insoluble PEC are much less defined than the structure and properties of soluble PEC. In insoluble PEC, more or less extended regions with interchain ion pairs co-exist with separate loops and 'tails', which can be regarded as structural defects. The higher the conversion θ the higher is the number of 'paired' units in PEC. The free hydrophilic units localized in loops and 'tails' determine the degree of PEC swelling in solvents [19].

If Z = 1 but $\theta < 1$ (incomplete reaction), water-soluble PECs are formed as equilibrium products of complexation. The units of the starting polyelectrolytes not involved in the formation of interchain ionic bonds play the role of lyophilic (hydrophilic) fragments of macromolecules due to which PEC particles are held in solution. When $\theta \rightarrow 1$, water-soluble PECs are formed if the molar ratio between the units (φ) of the polyelectrolyte taken in deficiency (blocking polyelectrolyte) and polyelectrolyte taken in excess (lyophilizing polyelectrolyte, LPE) in PEC is smaller than 1, i.e., if the LPE is present in excess in the system. In the work [20], these pair-forming polyelectrolytes are called guest polyelectrolyte (GPE) and host polyelectrolyte (HPE), respectively. Below we use these abbreviations.

As a rule, reactions of polyelectrolytes can form equilibrium systems in which soluble and insoluble PEC co-exist. The equilibrium can be shifted toward the soluble PEC by introducing an excess of

HPE as the lyophilizing high-molecular-mass component in the system.

If the degree of polymerization of HPE is not lower than that of GPE, soluble products are formed. The critical value (φ_c) is the limiting degree of occupation of GPE chains in HPE above which the hydrophilicity of the excess single-stranded segments is not sufficient for PEC particles to be held in solution; as a consequence, phase separation occurs in the system [19]. For the majority of defined systems, φ_c is 0.2 - 0.6 and depends on the chemical nature of polyelectrolytes. Soluble equilibrium PEC with $0 < \varphi < \varphi_c$ behave in dilute solutions as if they were individual polyelectrolytes with different charge densities.

The behavior of non-stoichiometric PEC formed by the linear polyions of HPE with a high degree of polymerization and small charged GPE particles (cationic dendrimers, globular proteins, etc.) in solution is similar to the behavior of non-stoichiometric PEC consisting of linear polyelectrolytes. The non-stoichiometric PEC containing hydrophilic single-chain segments of HPE and the relatively hydrophobic GPE - HPE double-chain domains are regarded as some kind of block copolymers [21].

Yet another family of soluble PECs are the so-called block ionomer complexes (BIC), including a block copolymer of ionic and non-ionic blocks as one of the components [22-26]. These complexes remain soluble even when $Z = \varphi = 1$. For example, a BIC based on the diblock co-polymer of polymethacrylate - block polyethylene oxide and polyvinylpyrrolidone (PVP) is known.[24] For $Z = \varphi = 1$, the BIC unimer contains a hydrophobic domain of the stoichiometric polymethacrylate - PVP complex covalently bound with the hydrophilic block of polyethylene oxide and is thus an amphiphilic macromolecule. The behavior of these compounds is similar to that of surfactants. As in the latter case, polymer micelles are formed at concentrations higher than the critical micelle concentration. A micelle has a water-insoluble nucleus (PEC block) and a hydrophilic 'crown' (polyethylene oxide block), due to which it is soluble in water and has aggregation stability. Unlike classical polymer micelles, they disaggregate when low-molecular-mass salts are added because of nucleus decomposition caused by the electrostatic screening of the charge.

The above tendencies in the formation and behavior of PEC are characteristic of systems in which the Coulomb interactions are dominant. More complex behavior is shown by PEC where electrostatic interactions co-exist with hydrophobic, donor - acceptor and dipole - dipole interactions and hydrogen bonding. The regularities of complexation of these multifactor polyelectrolyte systems are not discussed here.

Major attention in this chapter is paid to structure organization, physicochemical properties and applications of chitosan PECs with different polyanions.

Chapter 3

POLYELECTROLYTE COMPLEXES OF CHITOSAN WITH BIOPOLYELECTROLYTES AND MODIFIED NATURAL POLYANIONS

1. INTERPOLYELECTROLYTE COMPLEXES OF CHITOSAN WITH POLYSACCARIDES

The majority of publications are devoted to studies of chitosan polycomplexes with polysaccharides of another nature. All used polysaccharides are conventionally divided into four groups: plant, animal, bacterial polysaccharides, and synthetic polysaccharides. The typical representatives of plant polysaccharides are cellulose, hemicelluloses, xylans, glucomannans, galactans, pectins, starch, and algal polysaccharides (agar, carrageenan, fucan, galactans, etc.); animal polysaccharides include chitin, glycogen, and glycosaminoglycans of connective tissue (heparin, hyaluronic acid, chondroitin sulfates and keratan sulfates); the well-known polysaccharides of microorganisms include dextrans and mannans, hyaluronic acid from streptococci A and C, etc.

Polysaccharides exhibit high biological activities; at the same time, they are low-toxic, biocompatible and biodegradable. Due to this combination of properties, they have many useful scientific and practical applications; in particular, they are used for constructing various biomaterials by complexation with oppositely charged biomolecules.

1.1. Interpolyelectrolyte Complexes of Chitosan with Animal Polysaccarides and Modified Animal Polysaccarides

The best studied chitosan PEC include complexes with animal polysaccharides, including those with glycosaminoglycans (GAG). As it is known GAGs play an important role in living organisms. Thus, the mucopolysaccharide keratan sulfate is part of the proteoglycan complex, which can form complexes with hyaluronic acid and is involved in the formation of blood vessel walls [27, 28]. Heparin can accelerate multiplication of some cells and inhibit growth of others, it stabilize the growth factor; moreover, it possesses anticoagulant, anti-sclerotic and anti-inflammatory properties and is one of the most widely used medicines for treatment and prevention of cardiovascular diseases. Chitosan, in turn, is a biocompatible and bioresorptive polymer; it is cytocompatible with keratinocytes and fibroblasts the adhesion and proliferation of which on chitosan depends on DD (the higher the DD the higher the cell adhesion). Fibroblasts adhere better than keratinocytes, but proliferation of the latter is faster when the DD of chitosan is higher [29]. *In vivo*, chitosan can be bound to GAG (in particular, to chondroitin sulfate and hyaluronic acid) [27] to form stable PEC.

The PECs of chitosan with hyaluronic acid, chondroitin sulfate, chondroitin 6-sulfate, heparin and sulfated cellulose and methods of their analysis were described [30-40]. It was found that the stoichiometric PEC of glycol chitosan with hyaluronic acid [31] could be obtained at pH 6.0. In the structure of chitosan, there is one weakly basic amino group per pyranose ring; in the structure of hyaluronic acid, there is one weakly acidic carboxyl group per two pyranose rings (i.e., the number of amino groups is not equivalent to the number of carboxyl groups). In addition, there is steric hindrance for the rotation of pyranose rings, which rationalizes the deviation from the stoichiometric composition at low pH values. In contrast to hyaluronic acid, the heparin molecule has two strongly acidic (sulfate) and one weakly acidic (carboxyl) groups per two pyranose rings; therefore, in the case of the chitosan PEC formed at low pH values, the calculated and experimental data were in agreement [31].

The physicochemical aspects of the interactions of highly deacetylated samples of chitosan (DD 98%) with chondroitin 4-sulfates (55% and 70% sulfo groups, molecular mass 45.5 kDa), chondroitin 6-

sulfate (90% sulfo groups, 59 kDa) and hyaluronic acid were considered [32,33]. As could be expected, the degree of electrostatic binding depended on the degree of ionization of polyelectrolytes, the nature of ionogenic groups and charge density.

For chitosan PEC with chondroitin 4- or 6-sulfate, the result was independent of the position of the sulfo group in the polysaccharide units. When the salt was formed (sodium) of chondroitin sulfate was mixed with chitosan, maximum precipitation was observed at a molar ratio (ρ) of 1 between the glucosamine residues and the residues of the sulfo and carboxylic groups. The results agreed with previously published data [30, 31]. For the acidic form of chondroitin sulfate, the equivalence point corresponded to ρ = 0.75; in this case, the NH_3 groups of chitosan formed ionic bonds with all sulfo groups of chondroitin sulfate (which completely dissociated) and only half of all carboxylic groups. Stable polyelectrolyte complexes of chondroitin sulfate – chitosan (potential colon-targeted drug carriers) [34] were formed at different pH values with various ratios or at a fixed molar ratio ρ = 0.5 under various pH conditions and quantitated with the use of solid-state 13-C CP MAS NMR and element analysis. FTIR and 13-C NMR clearly showed H-bond formation at low pH, indicating that in addition to the electrostatic interaction, H-bonding may be involved in complex formation. Chondroitin sulfate – chitosan complex yield [34], FTIR spectra [34], thermogravimetric [34 -36] and dielectric [36] analysis were used to study the degree of interactive strength between polyions.

Complexation of chitosan with both salt and acid forms of hyaluronic acid led to the formation of a 1: 1 PEC [32]. The effect of the DD of chitosan on the stability of the resulting complexes with hyaluronate showed that as the DD (charge density) increased, the stability of PEC, as well as the cooperativity of the electrostaic interactions, increased [33]. In the course of the formation of chitosan PEC with acid forms of GAG, the carboxylic residues of polyelectrolytes were deprotonated. Depending on the pK_a value, deprotonation was either partial (for chondroitin sulfate) or complete (for hyaluronic acid, whose pK_a is lower) [32]. The order of the addition of polyelectrolytes did not affect the complexes formed. PEC characterized by high stability over wide ranges of pH and ionic strength were obtained by the interaction of chitosan with GAG in both acid and salt forms.

Recently nanosized particles (NPs) of GAG-based PEC were described [37-39]. The one-shot addition of polycation and polyanion solutions (ρ = 0.09-19.2) used for NPs permitted formation of both cationic and anionic particles from both polysaccharide pairs [38]. For most conditions studied, colloidally-stable, nonstoichiometric NPs were formed in solution. However, NPs formation was inhibited by flocculation at charge mixing ratios near 1. When adsorbed to surfaces and dried, some formulations resulted in discrete NPs, while others partially or completely aggregated or coalesced, leading to different surface morphologies. Nanosized (average diameter 162.8 ± 18.9 nm) and stable chitisan – *g*-PEG/heparin PEC micelles were produced by a self-assembly process [38]. PEC micelles facilitated the intracellular delivery of heparin, triggered the caspase activation, and consequently promoted apoptotic death of cancer (mouse melanoma 13 16F10). In the work [39] self-assembled enoxaparin (low molecular weight heparin) PECs were formed with chitisan and chitosan derivatives (CS-cystein, trimethyl CS (TMC), PEGylated TMC copolymers). Soluble NPs in the size range of 200-500 nm with spherical morphology could be obtained in the pH range of 3.0 - 6.5, with positive charge and drug encapsulation efficiency of approximately 90%. The PECs of chitosan derivatives were rather stable against ionic strength even when ionic strength was as high as 300 mM, as a consequence of the steric effect of methyl groups, polyethylene glycol segments and the formation of disulfide bonds for TMC, PEGylated TMC and chitosan-cystein conjugates, respectively.

Complexation of heparin, chondroitin sulfate A, chondroitin sulfate C with N-acylated chitosan derivatives (N-benzoylchitosan, N-myristoylchitosan, N-propanoylchitosan, etc.) with different degrees of substitution is reported [41]. It was noted that maximum turbidity of the system increased with the degree of acylation of chitosan and decreased to minimum because of PEC precipitation after the system had been stored for 24 h. It was shown that at pH 2.8 and 4.5 stoichiometric PEC of chondroitin sulfates A and C with chitosan and N-benzoylchitosan could be formed.

Materials based on biodegradable polymers such as chitosan and GAG possesses useful biological properties. They are cytocompatible and can stimulate wound healing processes and induce cell proliferation for the recovery of cartilage, skin tissue, etc. Novel mucoadhesive nasal inserts based on chitosan/hyaluronate polyelectrolyte complexes for vancomycin or insulin delivery were obtained at different pH and molar ratios [35]. Chitosan beads with encapsulated gentamicin sulfate were used as templates for layer-by-layer assemble with hyaloronic acid [40]. These beads showed slow sustained drag release *in vitro* and could be used as a drug delivery system.

Hydrolysis of complexes with specific enzymes (chondroitinase, hyaluronidase) and the ability of biomaterials based on chitosan and GAG to be cytocompatible with keratinocytes and chondrocytes were studied [27, 42]. It was shown that chitosan in PEC acted as a protector against hydrolytic cleavage of GAG. However, chitosan showed better results in experiments on cell adhesion and proliferation and in the series of *in vivo* experiments, which demonstrated its higher efficiency as a wound healing agent compared with the efficiency of a PEC of chitosan with GAG.

Films of chitosan – hyaluronic acid PEC in NaCl solutions can reversibly change their form under the action of an electric field [43]; the bending angle can be adjusted by varying the ionic strength of the salt solution and the electric field strength. These 'electrosensitive' PECs are promising for use as the components of artificial organs (e.g., as contractive structures mimicking the muscle work), sensors, on-off switches, etc.

Porous constructions [44] (membranes, blocks, tubes, etc.) with controllable (by varying the freezing - drying conditions) pore sizes (1 - 250μm) were obtained on the basis of chitosan. These structures were later used for the formation of chitosan - GAG (heparin) PECs. Complexation did not violate the inner microstructure of pores.

Glycosaminoglycans such as heparin and heparane sulfate are capable of binding the growth factors and cytokines and also initiate or enhance their synthesis. Thus, constructions containing chitosan – GAG complexes can be used as systems capable of retaining and accumulating the necessary factors secreted by the colonizing cells and even growth factors from liquid around tissues. These constructions are suggested as potential implants in tissue engineering [44-46].

Highly selective membranes were obtained from films of the PEC of chitosan (550 kDa, DD 87%) with carboxymethylchitin (CMCS, 60 kDa, 1.17 carboxymethyl groups per unit) possessing a specific structure stabilized by a system of ionic and hydrogen bonds; the membranes could be used for separating water - organic mixtures and for hemodialysis [47]. The composition of CS–CMCS PEC used for casting of films varied from 0.5 to 4.0 (CS: CMCS); this determined the conditions of additional hydrothermal treatment and the basic physicochemical characteristics of films such as swelling, sorption capacity, mechanical and transport properties, etc. At higher chitosan contents in films, the degree of swelling in water increased drastically. Films with a stoichiometric composition had the lowest degree of swelling. Thermal treatment of films led to dehydration of interchain salt bonds and resulted in covalent cross-links (amide bonds). The resulting partially cross-linked PEC films were characterized by poorer swelling in all solvents and lost solubility in acidic solvents (in contrast to thermally non-processed PEC).

The system of ionic and hydrogen bonds in CS–CMCS polyelectrolyte complexes ensured the formation of a more rigid (compared with the chitosan structure), strained and fragile structure. When the composition of the PEC film approached stoichiometric, their sorption capacity decreased, while the stability and rigidity slightly increased. Thermally treated stoichiometric PECs having greater numbers of ionic and covalent cross-links than non-stoichiometric PECs are demonstrated the highest selectivity in the pervaporative separation of water - organic mixtures, and high permeability. It was noted, however, that CS–CMCS PEC films are inferior to chitosan – poly(acrylic acid) (CS–PAA) PEC films in ability to separate water-enriched water - propan-2-ol mixtures [48, 49]. This effect was explained [47] by the presence of two bulky substituents in the CMCS macromolecule because of which the structure of CS–CMCS PEC probably had lower packing density than the structure of CS–PAA PEC, despite the higher steric complementarity of co-polyelectrolytes in the case of chitosan and CMCS.

In the work [50] investigated the elaboration of nanoparticles from the pH-induced self-complexation of the amphoteric polysaccharide N-sulfated chitosan. The acidification of aqueous solutions of chitosan having a degree of acetylation of 24% and a degree of sulfation of 34% or 56% was followed stepwise by turbidimetry, dynamic light scattering, and electrophoresis. With the highest sulfated chitosan, no turbidity was recorded between pH = 7.8 and 2.0, traducing a high apparent solubility of the polymer chains in this domain of pH. With the lowest sulfated chitosan, a steady increase in turbidity was monitored from pH = 6.90 to 6.15 followed by the flocculation of the polymer at pH approximately 6.0. In this range of pH, the polymer phase separated to yield particles having hydrodynamic diameters decreasing from 350 to 260 nm and an almost constant negative charge. These particles were assembled by electrostatic interactions between the protonated amino residues and the sulfate functions and stabilized by an excess of surface sulfate groups. The particles could be separated from the reaction medium and concentrated by centrifugation-redispersion cycles without alteration of their structure.

The two derivatives of chitosan were used to prepare stable multilayer polyelectrolyte films through a layer-by-layer (LbL) deposition technique [51]. Low molecular weight chitosan was modified by glycidyltrimethylammonium chloride to obtain the cationic form of the polymer. The anionic form of chitosan was obtained by sulfonation of carboxymethylchitosan with trimethylamine - sulfur trioxide. The films revealed smooth surfaces and linear growth of the thickness during LbL adsorption as measured by atomic force microscope. Contact angle measurements revealed a very hydrophilic nature of the formed films. These films thanks to biocompatibility and bacteriostatic properties of chitosan and its derivatives may be potentially useful for many biomedical and environmental protection applications.

Composite of carbonate-containing low-cristallinity nanoparticle hydroxyapatite (HA) and chitisan – phosphorilated chitosan PEC was evaluated *in vitro* and *in vivo.* The HA-PEC nanocomposite with complicated poros structure was prepared by a biomimetic method. The biodegradable composite promoted osteoblast adhesion, proliferation [52].

1.2. Interpolyelectrolyte Complexes of Chitosan with Plant Polysaccarides and Modified Plant Polysaccarides

Chitosan PEC with carrageenans (anionic linear sulfated polysaccharides containing D-galactopyranose residues linked by regularly alternating α $(1 \rightarrow 3)$ and β $(1 \rightarrow 4)$ bonds) have been considered [53-57]. The ι- and κ-carrageenan molecules include 3,6-anhydro-D-galactopyranose residues and have a double helix conformation in aqueous solutions, which determines their gel-forming ability. λ- Carrageenan has a high content of sulfo groups and no anhydrogalactose residues; in addition, it cannot form gels. Carrageenans are non-toxic and biocompatible and can therefore be useful in food and pharmacological industries.

When chitosan forms complexes with κ-, ι-, and λ-carrageenans, the type of the resulting PEC is dictated by the charge density on polysaccharide molecules, their concentration, and conformation (globular or helical) of carrageenan macromolecules [57]. Chitosan PEC with ι-, κ-, and λ-carrageenans [54] were prepared for medical and biological applications. It was noted that at concentrations higher than 0.1% for aqueous chitosan and higher than 0.3% for λ-carrageenan, PEC were formed as gels in the bulk; insoluble PEC were formed at lower polymer concentrations. Complexes can be obtained from concentrated (4%) and highly concentrated (20%) polysaccharide systems, their mixing generally led to the precipitation of complexes [55-57].

Structural organization of gels and consequently their mechanical and rheological properties depended on the chitosan content in PEC and on the nature of carrageenans [54]. λ-Carrageenan did not substantially affect the properties of PEC; it played an auxiliary role of a co- polyelectrolyte required for the preparation of PEC with chitosan. ι- And κ-carrageenans were more important for the complexation; their gels with chitosan were thermally sensitive, which was explained by the thermally reversible helix - knot conformation transition of carrageenan molecules, occurring as a sol - gel transition at low temperatures.

The least viscous gels were prepared by using λ-carrageenan. Static viscosity increased substantially on passing to ι- and κ-carrageenans. As a result, chitosan gels with κ-carrageenan were more rigid. The reason for these distinctions is possibly the formation of additional cross-links, which formed in systems with ι- and κ-carrageenans, but were absent in the case of λ-carrageenan.

pH-Sensitive polyelectrolyte complex hydrogel systems of chitosan with carboxymethylcellulose (CMC) and κ-carrageenan were described [53]. The homogeneity and swelling ability of these hydrogels depended on the chitosan: carrageenan: CMC ratio and the content of low-molecular-mass salts in polymer solutions. The threshold polyelectrolyteto salt ratios were determined at which homogeneous or heterogeneous gels formed or not. The degree of swelling of chitosan – carrageenan – CMC tripolymer gels decreased proportionally to the decreasing content of carrageenan in PEC because of the decrease in the osmotic pressure and reached maximum at pH 11 - 12; a similar dependence was also observed for chitosan – carrageenan bipolymer PEC.

Direct mixing of chitosan and alginate solutions would readily coagulate or form gels, leading to unfeasibility in the production of defined form of PEC particles or fibers. Alginic acid – linear copolymer of homopolymeric blocks of mannouronat and its epimer guluronate residues, covalently linked together. Alginates are extracted mainly from seaweeds (*Laminaria*) or bacteria's.

Alginate – chitosan beads [58, 59] or microparticles [60] were synthesized by chitosan treating of calcium alginate particles. Chitosan – alginate beads were prepared by dropping chitosan solution into a sodium alginate solution without any salt [61, 62]. Alginate – chitosan – alginate microcapsules were used as implantat in the subcutaneous space of mice [63]. Proper selection of the reaction pH, polymer concentration and hence charge density, and hardening time was important and determines the characteristics of PEC particles. Based on alginate – chitosan PEC fibers [64], fibrous scaffolds [65] were described. Gelation problem associated with the direct production of alginate – chitosan hybridized fibers has been overcome by the use of chitosan in the form of the chitosan – citrate emulsion, which had been prepared from olive oil and sodium dodecyl sulfate aqueous solution [64]. The chitosan – citrate emulsion was mixed with the alginate aqueous solution to obtain the spinning dope suspension for the preparation into alginate – chitosan hybridized fibers by wet spinning. The results showed that the tensile property values were greatest at the lowest content of incorporated chitosan (0.5% w/w). But it was shown that absorption of the Amido Black 10B dye (model of anionic drug) in the fibers increased with an increase in the chitosan content. Alginate – chitosan PEC- based films [66-68], membranes [69-70], hydrogel coatings on silicon wafers or paper [71] and titanium surface [72] were obtained. For preparation of good film, it is desirable that the polyelectrolyte complex coacervate suspension, which is cast to form the film, must be fine and uniform. The reaction rate of chitosan to alginate is very high, it can be controlled by addition water miscible solvent like acetone and methanol [66]; propylene glycol (0.5%) was used to obtain flexible film and for good peeling of film. Besides precipitate method, layer-by-layer technique was used for film formation [65].

PEC of chitosan with its sulfated derivative [73] (its biological properties are similar to those of heparin) and alginic acid [74, 75] were obtained. The effects of the DD and molecular mass of chitosan on the stability of the complexes formed were studied. The higher the DD and the lower the molecular mass, the more stable the complexes. Yan et al. [75] studied the PEC of sodium alginate (100 kDa) with chitosan samples (DD 95%) with low (low-molecular-mass chitosan, LM CS, 130 kDa), medium (530 kDa), and high (HM CS, 1000 kDa) molecular mass. The HM CS complex with alginate formed large gel-like coacervates, while the LM CS – alginate polyelectrolyte complexes were more finely dispersed. The most homogeneous films with the highest water permeability in neutral media were obtained from PEC of this type.

Interactions in chitosan – pectin (plant polyanionic saccaride) systems and the properties of the PEC obtained were studied in many works [76-82]. The binding parameters of LMW chitosan to pectin in solutions and multi-layers were evaluated by Hill's equation and Karisson's model respectively [81]. In both systems, the binding of chitosan to pectin was anti-cooperative, and the binding constant was 196 ± 8. Membranes and fungistatic laminated films based on chitosan – pectin PEC [79, 80] possessed pervaporative properties. The swelling ability of the films of these PEC [78] can be controlled by varying the composition and pH of the medium; the highest values were obtained at pH < 2 and >7. Chitosan - pectin complex prepared in molar ratio 1:9 showed the highest mucoadhesive properties and pH-dependent swelling sensitivity suitable for colon delivery of vancomycin [82].

Complexation of chitosan (230 kDa, DD 80%) with polygalacturonic acid (main component of pectin) (PGA, 102 kDa, 87.5% galacturonic acid) was studied by conductometry [83]. The conductivity of the system was measured during titration of sodium polygalacturonate with solutions of chitosan hydrochloride; the change in the slope of the conductometric curve suggested complete PEC formation. A stoichiometric PEC formed at the equivalence point (CS: PGA = 0.98 ± 0.04), which was confirmed by several independent methods (viscosimetry, potentiometry, turbidimetry, and gravimetry); the position of the equivalence point was independent of the order of mixing of polyelectrolytes. The conversion of polyelectrolytes (degree of complexation) was determined, which depended on the molar ratio of the reactants. The dependences found suggested that under the chosen conditions complexation in the CS–PGA system was cooperative.

Stoichiometric complexes of chitosan with carboxymethylcellulose, alginic and polygalacturonic acids were obtained at equimolar ratios of components [84-87]; the highest degree of swelling of PEC was observed at pH 4.0 - 5.8 [87]. Polyelectrolyte complexes of chitosan with carboxymethylcellulose revealed clot-inhibition *in vitro* [88], were used for pulp tissue regeneration [89], based on such PEC hydrogels were pH-sensitive and undergo enzymatic degradation [90].

Chitosan (370 kDa, DD 82%) PECs with carboxymethylated cashew gum (*Anacardium occidentale*) [91] and carboxymethylglucomannan extracted from tubers of *Amorphophallus konjac* were described [92]. The highest product yield was achieved at a ratio of chitosan: cashew gum of 25: 75 [91]. Thermogravimetric and IR spectrometric analyses of the PEC were performed and the activation energies of thermal destruction and the degradation temperatures of polysaccharides and their complexes were determined. It was found that the thermal stability of PEC decreased relative to that of the starting polysaccharides. Based on chitosan – carboxymethylglucomannan PEC [92], nano- and microparticles were obtained, their size varied from 50 nm to 1.2 μm depending on the initial concentrations of the polysaccharides, pH and ionic strength of solutions. It was shown that nanoparticles based on chitosan – carboxymethyl glucomannan PEC could effectively be used as systems for encapsulating water-soluble drugs.

Polyelectrolyte complexes of chitosan and gum kondagogu (*Cochlospermum gossypium*), free exudate gum that belong to the family Bixacoae, were prepared by mixing polymeric solutions of concentrations 0.02-0.18% w/v. The complexes formed were loaded with diclofenac sodium. PECs showed increase in relative bioavailability compared to the free drug when administered orally to the rats [93].

Polyelectrolyte complexes of chitosan with plant polysaccharides (alginic acid, pectin, carrageenan and so on) are extensively used for the creation of systems with controllable release of drugs [58-60, 66, 82, 92-106], for stability improvement of enzymes [61], as materials for wound healing [66, 71, 107], for cells incapsulating [72], as biocompatible materials and biological construct for tissue engineering applications [63, 65] were used.

In the work [65] the blood compatibility of the chitosan – alginate PEC fibrous scaffold could be significantly improved by incorporation a small amount of heparin in the polyelectrolyte solution during fiber formation. The platelet production and platelet adhesion on the chitosan – alginate – heparin fibrous scaffold were comparable to those on the resting control. *In vitro* cytotoxicity test showed that the scaffold was not toxic to human mesenchymal stem cells, and *in vivo* (rats) scaffold was biocompatible

Immobilized indometacin was obtained from genipin – cross-linked chitosan – sodium alginate PEC [105]. The swelling ability of these systems decreased with pH and alginate concentration in the gel-forming solution. Moreover, this parameter also depended on the degree of cross-linking. As could be expected, the degree of cross-linking decreased with the increasing degree of protonation of the amino groups of chitosan because the nucleophilic attack of the dihydropyran ring of genipin was inhibited. As a result, the rate of indometacin release from PEC particles with low degrees of cross-linking (formed at lower pH) was higher.

Films [101], microcapsules [102], and pellets [96, 103, 104] of chitosan – carrageenan complexes were prepared for the starting concentrations of carrageenan of 0.1% - 4.0% in solutions. One of the most important characteristics of these systems is the swelling ability, depending on structure of PEC, pH of the media, and determines the degree of release of drugs. The swelling ability of drug matrices based on chitosan – alginate and chitosan – carrageenan PEC was analyzed using the Hopfenberg model [96] for evaluating the contributions of various physicochemical processes (diffusion, relaxation, sorption - penetration, etc.). Because of the better water-binding ability of carrageenans (e.g., compared with the same ability of hydroxypropylmethylcellulose [104] or alginate [96] for the same polyanion content in PEC), the pelletized forms were liable to undergo fast (in less than 30 s) disintegration after the swelling stage.

Thus, chitosan – carrageenan systems proved less favorable matrices for the preparation of drugs with a prolonged activity than chitosan – alginate PEC [96].

1.3. Interpolyelectrolyte Complexes of Chitosan with Microbial Polysaccarides and Modified Microbial Polysaccarides

A study of chitosan PEC with sodium dextran sulfate (SDS) was reported [108-112]. The PECs obtained were sensitive to changes in pH and the solution ionic strength [108]. At pH 10.5, the swelling was maximum and decreased when the solution ionic strength increased. Variation of pH, molecular mass and the rate and order of mixing of polyelectrolytes led to PEC with a molar ratio of CS: SDS from 0.8 to 2.5 [109, 110]. Two types of polyelectrolyte complexes [110] with different molecular structures and properties could be formed; In the first series (A), PECs were obtained by mixing solutions of chitosan and SDS of the same concentration (1 g/ litre) in 2% acetic acid at volume ratios of 4 : 1 and 1 : 4; in the second series (B), equal volumes of polymer solutions of the same concentration were mixed in 1% hydrochloric acid. As a result, PECs of the series A were characterized by a greater number of salt bonds between $–NH_3^+$ and $–OSO_3^-$ groups than PECs of the series B. Moreover, hematologic tests showed that PEC of the series A had anticoagulant (antithrombogenic) properties, while PEC of the series B enhanced blood coagulation. Complexes with low-molecular-mass SDS inhibited hemocoagulation to a greater extent than PEC with high-molecular-mass SDS [109]. The order of mixing also produced a certain effect on the anticoagulant properties of PEC; polyelectrolyte complexes formed by the addition of chitosan to SDS had a more pronounced thrombogenic activity than complexes obtained by mixing the components in the reverse order.

The ability of the obtained PEC to bind lipoproteins and their anticoagulant activity were studied; a non-linear dependence of these parameters on the PEC composition was found. The CS–SDS polyelectrolyte complexes possess selective complexing ability with respect to low-density lipoproteins (LDL); the sorption selectivity decreased as the PEC: LDL ratio increased. The effects of the PEC concentration and PEC: heparin ratio on the interaction of PEC with heparin were follows: at PEC concentrations smaller than 10 mg/ml [111] and the same or lower content of heparin in blood plasma, the activity of heparin was enhanced; when the PEC concentration increased to 100 mg/ml, the antiheparin effect was observed. With the optimum ratio of the components, the CS–SDS complex showed higher anticoagulant and anticholesterol activities compared with those of the individual SDS. The effect of higher activity of SDS in its complex with chitosan was explained by the optimization of its physicochemical parameters (when SDS macromolecules were chemically modified with chitosan), in particular, by the decreased density of the negative charge and the flexibility of SDS macromolecules.

Kikuchi and Takewayashi [112] studied insoluble PECs of chitosan and carboxymethyldextran, their antithrombogenic properties depended on the pH of the polymer solutions; at pH 5.5 less thrombogenic products were obtained than at pH 3.5. These effects were explained by differences in the molecular structures of the PEC and the presence of residual carboxylic groups in them.

Thus, the efficiency of modification of anionic polysaccharides with chitosan through non-stoichiometric complexation and the possibility of controlling certain biological properties of PEC were shown.

Xanthan gum, a microbial exopolysaccharide consisting of a cellulossic backbone with two mannose and one glucuronic acid side chains on every second glucose residue, formed PEC gels with chitosan [113-118]. Molecular weight of xanthan can reach up to 6 millions Daltons, which makes it possible to create extremely viscous solutions at very low concentrations; xanthan is enzymatic resistant, stable over a wide range of temperatures and pH. Xanthan concentration was found to be the most critical parameter in PEC network formation. The hydrogel capsules were completely cross-linked at pH 6.2 [118] when initial xanthan solution concentration was at 1.5% (w/v). The swelling data also show that there is a significant decrease in the degree of swelling (SD) when pH is increased from 4.5 (SD = 2164) to 6.2 (SD = 1187), which can be the increase in the crosslinking density. pH-Sensitive swelling characteristics enable the controlled release of entrapped materials such as therapeutic agents, enzymes and bacteria [113-116], the targeted delivery and controlled release of encapsulated products for oral administration [117].

PECs were suggested for immobilization of enzymes. For example, chitosan – alginate PEC with immobilised urease [119] and ternary chitosan – polyanion – glucoamylase PECs [120] were described. SDS, polyvinyl sulfate and heparin [120] were used as polyanions. Insoluble PECs were prepared under certain conditions by co-precipitation of enzymes by mixing chitosan solutions with polyanions. The presence of an enzyme did not affect the composition of the resulting PEC. The enzyme activity depended on the type of PEC. Thus, in insoluble chitosan – SDS PEC, glucoamylase exhibited 100% activity, while in chitosan – polyvinyl sulfate and chitosan – heparin PEC, the activity decreased by 16% and 54%, respectively [120]. Immobilization of urease in the chitosan – alginate PEC led to increased thermal stability of the enzyme and stability on storage; after 70 days, the catalytic activity was 48% of the initial level, while native urease was deactivated by 50% in 7 days and by 100% in 20 days. The observed effects were explained by the appearance of a multipoint ionic interaction between the immobilized enzyme and PEC, which is the scaffold that imparts high conformation stability to the urease macromolecule [119].

1.4. Interpolyelectrolyte Complexes of Chitosan with Lipopolysaccarides

Yet another interesting type of chitosan complex is PEC with lipopolysaccharides (LPS), which are the components of the outer membrane of gram-negative bacteria. Lipopolysaccharides consist of regions differing in structure: lipid A having low structural variability, a highly variable O-specific polysaccharide chain and a less variable oligosaccharide segment (core). In rough mutants (R forms), there are no polysaccharide chains. Lipid A is the most conservative part of polysaccharides, which is invariable in structure in the majority of microorganisms studied. Lipopolysaccharides play an important role in the physiological functioning of bacteria; they are involved in transmembrane transport of various compounds and are endotoxins, antigens (the so-called O-antigens) and bacteriophage receptors, they are the primary target of the attack by antibacterial drugs and immunoglubulins. The polysaccharide chain of LPS is oriented toward the environment, and its fine structure is responsible for the recognition of bacteria and the specificity of immune response in higher animals and humans. Lipid A, which is closely associated with proteins, ensures the entity and stability of the cell membrane by decreasing its permeability and increasing its stability against hydrophobic agents; it is also responsible for the toxic properties of lipopolysaccharides [4]. The death of gram-negative bacteria as a result of inflammation, antibiotic activity, etc. liberates large amount of endotoxin. As is known, endotoxins (LPS, in particular, lipid A) have a pyrogenic action on the infected living organism; in particular, they result in severe intoxication (the so-called endotoxemy) and endotoxic shock [121]. Therefore, of particular interest is seeking the substances capable of reducing the toxicity of LPS.

LPS molecules bearing the negative charge on the surface (due to the presence of phosphate, pyrophosphate and carboxyl groups) interact with polycations (e.g., chitosan macromolecules) to form low-toxic chitosan – LPS complexes with different stoichiometry, which are promising as immunostimulants [122-125]. The mechanism of this complexation was studied by various techniques. It was found that the binding of chitosan to an endotoxin is a complicated process, largely depending on the macromolecular organization of endotoxins, the molecular mass and the DD of chitosan, reactant concentration and reaction temperature and time [122-125]. The complex is formed by means of not only ionic binding, but also other types of interaction. Stable chitosan – LPS complexes were obtained only after preliminary incubation of the components at 37°C. Presumably, the complexation was accompanied by additional dissociation of LPS aggregates and formation of a chitosan complex, which was quite stable over prolonged periods of time in solutions with high ionic strength [122-124]. An electron microscopic study of the complexes revealed a considerable effect of chitosan on the morphology of LPS. This effect manifested itself as the loss of the ultrastructural organization of the endotoxin and dissociation of aggregates formed by the granular PEC particles. It was noted that morphological inhomogeneity depended on the concentrations and ratio of the starting components and was probably related to the concentration dependence of the supramolecular structure of the LPS, which in turn affected their binding to chitosan.

It was found that chitosan specimens with lower molecular mass and degree of acetylation had higher affinity for LPS, which could be explained by the higher flexibility of chitosan macromolecules with a lower molecular mass. As a result, steric hindrance to the binding of high-molecular ligands diminished and hence the sites for binding became more accessible. The affinity for chitosan increased with the increase in the length of the O-specific polysaccharide chains of LPS; this enhanced the interaction between the components (increased the binding constants), but decreased the number of binding sites; this resulted in a decreased amount of bound LPS. As the LPS concentration in a solution decreased, the amount of the endotoxin capable of binding to chitosan increased. The stoichiometry of the complexes depended strongly on the LPS concentration in a solution [123-125].

In a series of biological experiments, it was shown that complexation of LPS with chitosan effectively decreased the acute toxicity of endotoxins [123, 124].

2. Polyelectrolyte Complexes of Chitosan with Nucleic Acids

One of the promising directions of research in biotechnology is the creation of non-viral vectors for transferring the specific genetic information to cells. These artificial viruses can successfully be used in gene therapy for the treatment of diseases caused by gene malfunction. The curing action is achieved by the expression of the gene vector, which replaces the defective gene or, vice versa, inhibits the excess activity, e.g., of the transforming gene or introduces a new gene for vaccination purposes.

To incorporate DNA into a cell, one can use [126] viral vectors, varieties of liposomal transport, specially designed protein or peptide vectors, including those addressed by the ligand - receptor or antigen - antibody interactions to certain cell lines, and vectors with positively charged HIV-TAT type peptides.

To obtain effective non-toxic and biocompatible DNA carriers, new systems are being developed that use polymers of natural origin, in particular, chitosan and its derivatives. In the resulting PEC, the positive charge of the amino groups of chitosan is neutralized by the negative charge on DNA; this leads to DNA compaction, protects the molecule from cleavage by endonucleases, and ensures gene penetration through cell membrane. Recent studies [127] on models of the membrane bi-layer showed that chitosan could penetrate into the cell breaking the double lipid layer of the cell membrane and could also induce the formation of transport channels in the latter, thus prolonging the supply of DNA and drugs to the cell.

Chitosan–DNA complexes has been used to prepare poros scaffold for tissue engineering [128], micro- [129] nanoparticles [130-134] as non-viral delivery vehicles or membrane [135].

Chitosan–DNA complexes were stable, possesses high porosity (more than 80%) [128]. PECs were not toxic to MG-63 osteoblast-like cells and caused only a mild tissue response when implanted subcutaneously in the backs of rats.

Chitosan/DNA-derived material surface properties such as charge, wettability, roughness have be studed to affect cell functions such as attachment, spreading, migration, proliferation, differentiation and aggregation.

Transfection of the chitosan - luciferase plasmid (pGL3-Luc) complex was studied by laser microscopy using FITC-labeled plasmid (FITC is fluorescein isothiocyanate) and Texas Red-labeled chitosan (Texas Red is sulforhodamine 101) [129]. The transfection activity and efficiency of incorporation in the cell were found to be affected by the molecular mass of chitosan, the stoichiometry of complexes and pH of the medium. The highest level of transfection of the chitosan - plasmid complex was observed when using chitosan with a molecular mass of 40 and 84 kDa, the ratio of chitosan nitrogen to DNA phosphorus (N : P) equal to 5, the serum content in the transfection medium 10% and pH 7. The study of the transfection mechanism showed that the chitosan – plasmid complex formed aggregates of the size 5 - 8μm, which were adsorbed on the surface of the cell, penetrated in it as a result of endocytosis and were further liberated from endosomes under the action of the lysosomal hydrolytic enzymes. Moreover, CS–DNA complexes in the form of nanosized particles (38 ± 4 nm [130]), 200 - 400 nm [131]) were proposed for the use in gene therapy. It was noted [131] that using chitosan with a molecular mass of 10 kDa formed large particles of the size 600 - 1000 nm. When the molecular mass of chitosan increased to 40 - 150 kDa, PEC particles of the size 200 - 400 nm could be obtained; in this range of molecular masses, the DD of chitosan and the N : P ratio did not appreciably affect the size of the aggregates.

For improving the release of DNA from CS–DNA complexes their internal structure was modified by incorporating a negatively charged poly(γ–glutamic acid) (γ–PGA) [132]. The analysis of small angle X-ray scattering results revealed that DNA and γ–PGA formed complexes with CS separately to yield two types of domains, leading to the formation of 'compounded NPs'. Such 'compounded NPs' mightily disintegrate into a number of small sub-particles after cellular internalization, thus improving the dissociation capacity of CS and DNA, increasing their transfection efficiency and cellular uptake.

Plasmid DNA (pDNA) and siRNA were encapsulated in CS nanoparticles (NPs) using a complex coacervation process [133]. It was shown that CS in CS–DNA NPs could delay the DNA release (pH 7.4) and protect the encapsulated DNA from nuclease degradation. Morphology of HeLa cells transfected *in vitro* with CS–siRNA complexes was studied using AFM. The results suggest that CS may be more capable than liposome (at used molecular weight, charge ratio and NA concentration in CS–NA NPs) in delivery siRNA to target cells.

Direct force measurements between small interfering RNA (siRNA) and chitosan molecules were utilize using force spectroscopy (AFM) [134]. To prepare a feasible model system for the chitosan – siRNA interaction, siRNA was immobilized to AFM tip, and chitosan was grafted onto gold coated surface via a linker of glutarealdehyde. The force measurements revealed that the adhesive interactions decreased in force strength and force frequency as the pH was increased from 4.1 to 6.1, 7.4 and 9.5, exhibiting distinct multimodal distribution of the interaction forces between siRNA and chitosan molecules at acidic pH and only negligible adhesive forces were observed at neutral or high pH. The strong pH dependence of siRNA–chitosan interaction can provide a convincing rationale for siRNA–chitosan complex formation and nanoparticle stability under low acidic conditions.

Three kinds of membranes for controlling cell spreading and aggregation were prepared by mixing chitosan and DNA at different ratios [135]. The membrane with the high ratio of chitosan had a less hydrophilic surface (contact angle measurement data). The surface of the polyelectrolyte complex membranes became rough as the DNA content increased (SEM observations). Human mesenchymal stem cells adhered and spread on membranes prepared at chitosan–DNA ratios of 1:1 and 3:1, while they did not on membranes prepared at chitosan–DNA ratio of 1:3. Cells aggregated on the membrane prepared at a chitosan–DNA ratio of 1:3. Cell viability was also higher on membranes prepared at chitosan–DNA ratios of 1:1 and 3:1 than that on the membrane prepared at a chitosan–DNA ratio of 1:3. The membrane with a high content of chitosan facilitated cell adhesion and spreading, while a high content of DNA suppressed cell adhesion and spreading.

Chitosan and polygalactosamine were used as non-viral vectors for the transfection of luciferase plasmid pGL3 in tumor cells [136]. Chitosan was several times more effective than polygalactosamine in supplying the plasmid complex to the target cells. The efficiency of the transfection depended on the stoichiometry of the chitosan - plasmid complex, pH of the culture medium and the molecular mass of chitosan. The optimum parameters were the chitosan: plasmid ratio 5: 1 and pH 6.9. The luciferase activity and gene expression efficiency for systems containing chitosan with molecular masses of 15 and 52 kDa were higher than for systems with chitosan with a molecular mass of 5100 kDa. The chitosan heptamer (1.3 kDa) did not show any gene expression.

Studies on DNA complexation with N-alkylated [137, 138] and non-modified chitosan (50 kDa, DD 99%) showed [137] that CS–DNA PEC formed at a charge ratio of 1: 1, while the hydrophobically modified CS–DNA complex formed at a ratio of 1: 4. The introduction of an alkyl substituent ($C_4 - C_{16}$) favored easier penetration of DNA through the cell membrane due to hydrophobic interactions and easier 'unpacking' of DNA from the carrier as a result of a weakening of electrostatic interactions between hydrophobically modified chitosan and DNA. The efficiency of transfection increased with the length of the hydrocarbon fragment in the range $C_4 - C_8$, where after its level changed insignificantly. While facilitating the penetration of plasmid into the cell, the hydrophic substituent of the resulting plasmid complex also performed the protecting function, namely, it protected the DNA molecule from the action of deoxyribonucleases. A methodological approach is now being developed that uses modification of chitosan with low-molecular-mass specific ligands to achieve the specific supply of DNA complexes to the target cells. Thus, artificial viruses become 'recognizable' by definite cells. For example, low-molecular-mass chitosan was modified [139] with galactose residues to make the vector 'recognized' by the liver cells having specific receptors for galactose. Galactosylated chitosan was used in vitro for the transfer of the reporter β-galactosidase in gene pSV to human hepatocarcinoma cells (HepG2, SMMC-7721) and hepatocytes of the L-02 line. The complex obtained possessed low cytotoxicity and high selectivity for hepatocytes, while the efficiency of gene transfection increased with the degree of chitosan galacto-sylation.

It is worthy to note the studies on the properties of liquid-crystalline complexes formed by chitosan with linear double-stranded molecules of nucleic acids (NA) of the B and A families and with synthetic double-stranded polyribonucleotides (PRN) [140-142]. In the reaction with NA molecules, chitosan converts the system into the liquid-crystalline state. Its form depends on the molecular mass of chitosan and the values of chitosan DD, pH and ionic strength of solutions.

The CS–NA complex was obtained by adding aqueous solutions of low-molecular-mass fractions of chitosan (2 - 31 kDa) with different contents of amino groups (46% - 85%) to the aqueous salt solutions of NA with stirring. Spectroscopy detected the formation of PEC dispersions. Circular dichroism (CD) studies of the dispersions showed that the interaction of chitosan with DNA resulted in the formation of liquid-crystalline dispersions with a characteristic helically twisted (cholesteric) spatial structure. The classical cholesterics formed by the phase exclusion of the double-stranded DNA molecules of the form B exhibited an abnormal negative band in the CD spectrum, which pointed to the levorotary form of the spatial cholesteric structure. The CD spectra of dispersions of the CS–DNA PEC showed an abnormal positive band, which corresponded to the cholesteric dextrorotary structure. PEC of chitosan and double-stranded DNA molecules of the form A, as well as 'pure' DNA molecules formed cholesteric dispersion with a levorotary spatial structure. Studies of the formation conditions of CS–NA PEC liquid-crystalline dispersions showed that a definite value of the molecular mass of chitosan is exist, starting from which the PEC formed in a cooperative way. The abnormal optical activity of CS–DNA and CS–PRN PEC dispersions depended differently on the molecular mass of chitosan. At pH near 6.8 (corresponding to the pK_a of chitosan in aqueous salt solutions), the optimum cholesteric packing of the CS–DNA PEC formed. Under the given conditions, the degree of ionization of chitosan was 50%. At pH < 7 and pH > 7, two types of CS–DNA liquid-crystalline dispersions formed, which differed in the molecular packing; dispersions of the first type were characterized by an abnormal positive band in their CD spectra, while dispersions of the second type did not show any changes in the CD spectra. For each chitosan sample with a definite DD, there exists a definite range of ionic strength values, within which the chitosa–DNA PEC with an abnormal band in the CD spectra can form. The liquid-crystalline dispersion was not formed outside these limits. Moreover, abnormal optical properties of the liquid crystals in question depended on the distance between the amino groups in chitosan molecules and the solution ionic strength; varying these two parameters affords liquid-crystalline dispersions that showed a negative or positive band in the CD spectra or did not show any abnormal optical activity at all.

To evaluate the steric and energy characteristics of chitosan complexation with DNA in solutions with variable ionic strength, a theoretical approach was developed, which allowed the construction of a mathematical model of binding of these polyelectrolytes that adequately described the experimental data [142]. The binding constant was calculated within this approach and was shown to decrease when the solution ionic strength increased. Note that treatment of the chitosan–DNA PEC dispersion with daunomycin (DAU), the molecules of which intercalate between the DNA base pairs, led to the formation of a cholesteric dispersion, in which the initial dextrorotary spatial structure changed to the levorotary structure. However, despite the reversal of the sign of the abnormal band in the CD spectrum, the X-ray parameters of the CS–DNA and CS–DAU–DNA PEC phases coincided; i.e., the introduction of DAU in PEC did not affect the distances between DNA molecules in the resulting dispersion.

The dependence of the optical properties of CS–DNA PEC liquid-crystalline dispersions on a number of parameters makes it possible to use the optical properties of these PECs as quality testers for chitosan samples employed in medicine and biotechnology. Studies of the mechanism of the transition of plasmid DNA to the liquid-crystalline state upon complexation with chitosan and the properties of the resulting PEC with different twist of the cholesteric structure for their future use for gene transfection are in progress.

3. POLYELECTROLYTE COMPLEXES OF CHITOSAN WITH PROTEINS

Association of protein and polysaccharide molecules through Coulomb and non-Coulomb interactions leads to the formation of PEC [143]. Under certain conditions, proteins (natural polyampholites) can show the properties of both polyacids and polybases; for complexation with chitosan, the protein should act as an acid.

The PECs of completely deacetylated chitosan with bovine atelocollagen were obtained [144, 145]. The structure of PEC in which co-polyelectrolytes were bound through electrostatic interactions was proved by various methods. It was noted that under the conditions chosen, gel formation of collagen competed with collagen complexation with chitosan. It was impossible to choose the conditions that provided exclusively the ionic complexation mechanism of the interaction of collagen with chitosan. At elevated temperatures at which protein denaturing was induced, it was possible to approach the synthesis of a stoichiometric polycation - polyanion complex. When chitosan was present in a large excess, the complexation was dominant, but there was hydrogen bonding between the protein and polysaccharide in addition to the electrostatic interaction. Biocompatible membranes were prepared from nanofibers of the chitosan PEC with collagen [146] cross-linked with glutaraldehyde and exceeded the widely used collagen sponges in their wound-healing properties.

Chitosan PEC with legumin of fodder beans and of its partial trypsinolysis product, legumin-T, were described [147, 148]. It was shown that the functional properties of plant proteins could be regulated by complexation with charged polysaccharides; complexation imparted solubility to the precipitated protein molecules near their isoelectric point, which allowed control over the solubility of precipitated globulins in weakly acidic and weakly basic media. Moreover, the formation of chitosan PEC with legumins considerably increased the stability of protein emulsions.

Chitisan–gelatin PECs are the perspective biomaterials. Gelatin is obtained through a controlled denature of the fibrous insoluble protein – collagen, which is the major component of skin, bone and connective tissue. It is characterized of no anti-genicity in comparision to collagen.

Chitosan–gelatin polyelectrolyte complex formation was described in work [149] with use of conductometric and pH titration. The chitosan interact with ampholitic gelatin (Gel) electrostatically and the interaction mainly depends on net negative charge on Gel macromolecular chains.

A combination of flexible protein and rigid polysaccharide results in biomimetic CS–Gel based biomaterials, drug delivary systems, surface modifiers, sponges and scaffolds for tissue engineering and non-viral vectors for gene therapy [150-160].

Porous scaffolds, mono- and bilayer biomimetic structures were developed on the basis of chitosan PEC with gelatin; due to their properties, they are recommended for use in tissue engineering [150-152,155,158, 159].

An asymmetric CS–Gel scaffold was prepared via controlling both the freezing temperature and heat transfer rate [150]. Bilayer scaffolds were fabricated with variance in the mean pore size from 30μm to 250μm. Scaffold was used as a temple for fibroblast and keratinocytes co-cultivation and provided suitable microenvironment for keratinocytes grow thing and differentiating. The presence of hyaluronic acid in scaffold enhanced its water retention ability and elongation. Such scaffold implanted subcutaneously degradated at 4th week *in vivo*.

In vitro it was shown [155] that gelatin-containing scaffolds had faster degradation rate and significant loss of material than CS in the presence of lysozyme. Mechanical properties of CS are affected by the addition of Gel although there was no clear trend. Three-dimensional CS and CS–Gel scaffolds supported fibroblast viability equally. However, CS membranes decreased cell-spreading areas, disrupted F-actin and localized FAK in the nucleus of HUVECs. Importantly, the lowest shear stress tested (4.5dyn/cm^2) for 3 h washed away cells on CS suggesting weak cell adhesion. No significant differences were observed in PECAM-1 expression.

Tubular sandwich structures based on chitosan – gelatin (4: 1 by mass) PEC with a wall thickness of 1mm and porosity of 81.2% were designed for tissue reconstruction of blood vessels [160]. It was shown that vascular smooth muscle cells of blood vessels spread and grew effectively on these polyelectrolyte structures.

Collagen-contained hybrid membranes [156] and scaffolds [161] were used with aim of investigation of cells adhesion, growth, proliferation and so on. The chitosan–collagen–gelatin (CCG) membranes with surface micro-patterns were fabricated by soft lithography [156]. The C3A cells on the CCG membranes with micro-patterns have higher metabolism and growth rates than those on the flat CCG membranes and on T-flask discs.

Tanabe et al. [162] prepared chitosan–keratin PEC films having enhanced mechanical properties (flexibility, elasticity and stability) compared with the properties of films from the starting polyelectrolytes and swelling in acid, neutral, and alkaline media (this was shown for pH 4.0, 6.3 and 8.9) due to the properties of the biopolymers of the PEC, namely, the acidic protein keratin having pI 4.9 − 6.1 and precipitating in acidic media and the basic polysaccharide chitosan having pK 6.3 and precipitating in alkaline media. Chitosan–keratin films have an antibacterial activity inherent in chitosan films, stimulate the attachment and proliferation of fibroblasts and serve as substrates for mammalian cells.

Self-assembled polyelectrolyte nanocomplexes (size range 200-500 nm) between methylated or PEGylated chitosan and insulin were prepared, and parameters influencing on complex formation were characterized [163]. Turbidimetric titration in combination with dynamic light scattering and laser Doppler anemometry were used to study the complexation process; the morphology of the PECs was determined using atomic force microscopy. Soluble PECs with spherical or sub-spherical morphology and smooth surface structure were obtained at pH 6.5−8.0 and MW of the polymers less or equal 25 kDa. PECs were stable and could protect insulin from degradation even at 50 ^{o} C for at least 6h. All complexes could be lyophilized without influencing the particle size and stability of insulin.

A method for selective removal of un-denatured β-lactoglobulin from cheese whey was described [164]. At pH 6.2, addition of 1.9 to 3.0 mg/ml of chitosan led to complete removal of β-lactoglobulin, ehereas at least 80% of the rest of whey proteins remained in solution. The production of cheese whey without β-lactoglobulin could help to expand the applications of dairy by-products in food processing and to isolate hypoallergenic whey protein concentrates.

The complexation of chitosan and mucin was described in works [165, 166].

Mucins representatives of a special subclass of high-molecular-mass glycoproteins with high carbohydrate content (up to 50% − 80% by mass). They possess branched oligosaccharide chains linked by the O-glycosidic bonds with the protein moiety of the macromolecule. Mucins are characterized by high content of such amino acids as serine, threonine and proline and such monosaccharides as D-galactose, L-fucose, N-acetylglucosamine and N-acetylgalactosamine as the main carbohydrates and sialic acids. The monosaccharide residues form oligosaccharide chains linked to the protein by the O-glycosidic bond of N-acetyl-galactosamine and the hydroxyl group of the side chain of serine or threonine [165].

The effects of the molecular mass of chitosan (70, 750, 2000 kDa), pH (2.9 – 5.0) and ionic strength (0.01 – 1.0M NaCl) of a medium on the complexation with bovine submaxillary mucin (400 kDa) containing 12% sialic acid residues were studied by fluorescence polarization [166]. Dextran sulfate or albumin was used as reference co-polyelectrolytes instead of mucin. For fluorescent polarization studies of the molecular mechanism of the interactions of chitosan with mucins, FITC-labeled chitosan (FITC-CS) was synthesized. For FITC-CS complexes with mucin, the degree of polarization increased with the mucin: chitosan molar ratio and the solution ionic strength. For example, at a ratio of mucin: FITC-CS = 19: 1 (750 kDa) and pH 2.9 (0.1M acetic acid), the degree of polarization increased by 61%. In addition, it was found that the increased molecular mass of chitosan, as well as the increased mucin content in the complex, also led to the increased degree of polarization. Thus, the number of sites capable of adding mucin increased with the molecular mass of chitosan. It was assumed that the chitosan molecule might have more than one mucin molecule bound to it, while binding of chitosan to dextran sulfate occurred according to the 'one-to-one' principle, and complexation of chitosan with albumin did not occur at all under these conditions. The maximum change in the degree of polarization in the interaction of chitosan with mucin was observed at lower pH (to pH 2.9 where the degree of ionization of chitosan was almost 100% and the degree of ionization of sialic acid residues was the least, 20%), which points to the dominant effect of chitosan on the interaction with mucin.

These studies allowed the researchers to conclude that the binding of two polymeric molecules involved not only electrostatic interaction between the D-glucosamine residues of chitosan and the sialic acid residues of mucin, but also hydrogen bonds and hydrophobic interactions.

Qaqish and Amiji [166] intended to use chitosan - mucin complexes for the creation of pH-sensitive drugs for localized supply of drugs to the alimentary canal.

Chapter 4

POLYELECTROLYTE COMPLEXES OF CHITOSAN WITH SYNTHETIC POLYANIONS

In the field of chitosan PEC with synthetic polyacid anions, the largest number of publications was devoted to chitosan PEC with polyacrylic acid (PAA) [48, 49, 167-185].

Insoluble PECs of chitosan (220 kDa) with PAA (202 kDa) of different compositions obtained at pH 3 - 6 were studied [168]. It was found that the solution ionic strength in the range 0.025 - 0.3 did not affect the formation of PEC. At pH < 3, insoluble chitosan–PAA PECs were not formed. The pH value affected the ionization of the reactant (charge density distribution along the chain) and hence the composition of PEC. The content of polysaccharide in the insoluble fractions of PEC increased with pH. Thus, PEC with controllable composition and properties could be prepared with consideration of the tendencies found for complexation.

The formation and structure of chitosan PEC (190 kDa, DD 80%) with PAA (500 kDa) was reported [169]. The dependences of the degree of complexation and the composition of PEC on pH were studied by potentiometric titration. It was shown that insoluble PEC enriched in PAA formed at low conversion. As the reaction proceeded further, PEC particles became enriched in chitosan; in neutral media, stoichiometric PECs with quantitative amounts of both polyelectrolytes were formed. In alkaline media, the complex decomposed with liberation of polyacrylate anions from PEC particles and isolation of chitosan-enriched insoluble PEC fraction.

The behavior of the PEC of water-soluble chitosan (DD 50%) with PAA was described [170]. Hydrodynamic studies showed that an increase in the chitosan fraction in PEC led to compaction of complex particles and hence to a decrease in the intrinsic viscosity of solutions; the minimum viscosity corresponded to the composition [CS]: ([CS] + [PAA]) = 1: 2. In acidic media (at pH < 2), PEC completely dissociated into the starting polyelectrolytes; in basic solutions (at pH > 8), optical transmission decreased from 100% to 70% because of partial precipitation of chitosan, which is characteristic of the behavior of water-soluble chitosan PEC. It was emphasized that hydrogen bonding also made a considerable contribution to complex stabilization along with electrostatic interactions, which played the major role in the formation of PEC.

One of the major applications of CS–PAA PEC is the reparation of semi-permeable membranes, which are widely used in food industry, medicine (in blood, plasma and blood substitute purification), pharmaceutical industry, waste treatment, oil products purification, etc. Semi-permeable membranes possess high separating ability (selectivity), high permeability, stability against the action of the medium being separated and sufficient mechanical stability. Note that semi-permeable membranes can serve as models of biological membranes characterized by high biological and chemical selectivity; therefore, their studies are also important for scientific purposes.

The structure of CS–PAA PEC's films, their mechanical and physicochemical properties, as well as the PECs potential applications as membranes for pervaporation, sensors, membranes for fuel cells, etc. were described [48, 49, 171-173, 181-183]. Studies of the mechanical properties of CS-PAA films showed [172] that tensile strength increased, while the relative lengthening during sample destruction decreased with respect to those of the starting polymer films. This result is explained by ionic cross-linking between the macromolecular chains in PEC. The liability of the PEC obtained at different polyelectrolyte ratios (depending on the pH of the medium) was evaluated from the change in the degree of swelling of films and IR spectroscopic data. Thus for CS: PAA = 1: 1 PEC, swelling increased at pH 3 and 9 and was minimum at pH 7. In the IR spectra of films after swelling at pH 3, the characteristic absorption band of the carboxyl group of PAA (1610 cm^{-1}) vanished; at pH 9, the band corresponding to the vibrations of the carboxylate ion of PAA was observed at 1590 cm^{-1}; and at pH 7, the absorption peak characteristic of the C=O stretching vibrations shifted because of the formation of the CS–PAA complex. The pervaporation properties of the obtained membranes were studied using various aqueous-organic mixtures; it was shown that the efficiency of pervaporation depended on the PEC composition [49, 172]. The sample films of chitosan (190 kDa, DD 80%) PEC with PAA (200 kDa) of different compositions (CS: PAA ratio 0.25 - 0.67) were obtained by casting from formic-acid solutions of polymers (pH 2). The properties of pervaporation membranes prepared, which were designed for separating water - organic mixtures were studied [49]. The mechanical stability of films was found to increase as PEC was enriched in rigid chains of chitosan. In contrast to PEC based on flexible-chain synthetic polymers, the films retained their stability in neutral and weakly alkaline media and demonstrated high transport and sorption characteristics. Materials with these properties are extensively used for the design of systems with controllable release of drugs in pharmacology, for the creation of constructions with definite chemo-mechanical properties [143, 174, 180-184], etc.

Wang et al. [173] obtained PEC by mixing equimolar amounts of an acetic acid solution of chitosan (300 kDa, DD 86%) with aqueous solution of PAA (230 kDa). After covalent cross-linking with glutaraldehyde, pH-sensitive membranes were obtained from PEC. A series of tests showed that the starting CS–PAA complex dissociated at pH < 2 and manifested high swelling at pH > 7. At the same time, the degree of swelling of CS–PAA membranes was high at pH < 2, decreased drastically as pH increased to 8 and then increased to a maximum at pH 11. Swelling of membranes in salt solutions increased with the ionic strength and depended on the valence of cations at constant values of the ionic strength; i.e., it was maximum in solutions of trivalent metal salts (Me = Al) and gradually decreased on passing to divalent (Me = Ca, Mg, etc.) and univalent (Me = K, Na) metals. Note that the mechanical properties of membranes became reversible when they were treated alternately with KCl and $CaCl_2$ solutions at equal ionic strengths; when membranes were treated with a $CaCl_2$ solution, their size increased, but diminished after treatment with KCl solution. This behavior is determined by the ion-exchange properties of PAA.

Of particular note are works where PECs were obtained by the free-radical polymerization of ionized monomers on a polymer matrix

[176-179]. For example, 'self-polymerizing' CS–PAA membranes were prepared using free-radical polymerization of acrylic acid on a chitosan matrix at moderate temperatures (37° C) [170]. The structure of polyelectrolyte membranes was confirmed by IR spectroscopy. The degree of swelling of membranes was found to depend on pH and composition; it increased with pH (from 5 to 10) and the PAA content in PEC. Thermogravimetric studies showed that interchain amide bonds formed in membranes at high temperatures (starting from 100° C). Subsequent thermal treatment at elevated temperatures led to further dehydration of ionic bonds between the chitosan and PAA macromolecules in PEC, mass loss and eventually PEC degradation.

CS–PAA composite nanofibrous membrane with 3D network nanostructure was prepared using an electrostatic process by adding succinic acid as branch promoter [181]. The smallest fiber diameter size distribution of the scaffold can be obtained when the PAA/CS ratio was in the range of 2:1 - 1:2 in a pH 3 environment. Negative charge nanofibers prepared using PAA/CS ratio 2:1 and pH 3 had an average diameter of 215 nm. The formation of the interconnecting 3D self-organized network structure can be build up with limited parasitic branching by crystallized siccinic acid. The results of sensing experiments indicated that the sensitivity of nanofibrous membrane-coated sensors was eight times higher that of continuous film-coated sensors. Sensitivity towards ammonia was 50 ppm at a relative humidity of 45%.

PEC CS–PAA was used as proton exchange ionically cross-linked membranes for fuel cells [182]. Membranes were characterized for morphology, their intermolecular interactions, thermal stability, and physico-mechanical properties using SEM, FTIR, DSC, sorption studies, and tensile testing, respectively. Among the blends synthesized, the membrane blend with 50 wt. % of CS and 50 wt. % of PAA, was identified as ideal for direct methanol fuel cell applications as it exhibited methanol permeability $3.9 \times 10^{-8} cm^2/s$, excellent physico-mechanical properties and comparatively high proton conductivity ($0.038\ S \times cm^{-1}$).

PEC CS–PAA membranes [183] and hydrogels [184] with high content of chitosan were used as drug release systems. Hydrogels with CS/PAA ratio 1:0.25 showed greater muco-adhesive property and maximum swelling in 0.1 N HCl [184] and complete release of amoxicillin and metronidazole in 10h. Permeability studies in relation of CS–PAA PEC membranes to a drug model (sodium sulphamerazine) were discussed in work [183]. Two membranes with CS/PAA ratio 95:5 were obtained: one prepared using PAA solution in 3.5% formic acid (CS–PAA 3.5) and another one using a PAA solution in 10% formic acid (CS–PAA 10). Water uptake results showed that PEC membranes were more hydrophilic than pure chitosan. Permeability values decreased with complex formation and were lower for CS–PAA 10 than CS–PAA 3.5. Diffusion coefficient values indicated that CS–PAA 3.5 had a higher drug retention capacity than CS-PAA 10.

For complexation with chitosan besides of proteins may be also involved synthetic polyamino acids [132, 186-189]. Poly(aspartic acid), recognized as green material, widely applied in agriculture, medicine, etc [188, 189]. Nonstoichiometric polyelectrolyte complex nanoparticles were prepared based on chitosan and polyaspartic acid sodium salt (PAsp). The physicochemical properties of the complexes were investigated by means of turbidity, dynamic light scattering, transmission electron microscopy and zeta potential. The results indicated that the slow drop-wise addition of chitosan solution into PAsp one allowed elaborate either anionic or cationic particles in the size range of 85-300 nm with proper CS and PAsp unit molar ratios; microstructure of the nanoparticles depended strongly on the unit molar ratio. Nanoparticles containing 5-fluorouracil were prepared by mixing and absorption method. *In vitro* and *in vivo* experiments indicated that drug-loaded nanoparticles presented a sustained release of drug compared to the 5-fluorouracil solution and the areas under curve were increased by about fout times. Polyelectrolyte complex hydrogel composed of chitosan and poly(γ-glutamic acid) was prepared for biological application, physical properties and cytocompatibility were investigated [186]; poly(γ-glutamic acid) – chitosan/DNA complex nanoparticles [132] were described in Part 2.

Other synthetic polyanions such as polyvinyl sulfate [120,190], polystyrene sulfonate [191, 192] and poly(2-acrylamido-2-methylpropansulfonic acid [193] were also introduced in complexes; the conditions influencing on complexes formation were discussed [190, 192]. The water-soluble nonstoichiometric polyelectrolyte complexes have been first prepared via the interaction of chitosan with excess of polystyrene sulfonate anions in acid media [192]. The region of existence of soluble complexes can be narrowed down through a decrease in the degree of polymerization of polyanion until fully degenerates as in the case of oligomeric anions.

In the work [193] it was found that enzymatic chitosan hydrolysis was not suppressed by combining it with poly(2-acrylamido-2-methylpropansulfonic acid. The results on the soil fungus *Trichoderma viride* immobilized on beads imply that chitosan, as well as polymer complexes are appropriate carriers for development of ecologically safe phytosanitary preparations.

Complexation of chitosan with tripolyphosphate (TPP) was described [194-196], and PECs were used for immobilization of enzymes such as lipases [194, 195]. Lipases are widely employed in biotechnology for separation of enantiomers, hydrolysis of fats to fatty acids, catalysis of cleavage of glycerol esters at the lipid - water interface, etc. The applications of these enzymes are limited by their instability and narrow range of the conditions (pH and temperature) of their biocatalytic activity. Lipases can destroy the outer layer of the microbial cell membranes. However, their shell contains waxlike lipids, and hence phagocytosis of bacteria cannot be affected by macrophages. One of the solutions to this problem is immobilization of lipases on polymer carriers capable of penetrating through the cell barrier. For this purpose, ionic gelation of chitosan solutions with tripolyphosphate was performed to obtain microcapsules based on the chitosan PEC used for immobilization of lipase from *Candida rugosa*. It appeared that the enzymatic activity of the lipase enclosed in chitosane microcapsules was fivetimes higher than the activity of the enzyme in alginate microcapsules. The molecular mass of the chitosan samples used for microencapsulation did not substantially affect the enzymatic activity of the enzyme. For maximum efficiency of lipase liberation from capsules, the DD of chitosan should be >90%.

It was demonstrated [196] that polyanion initiated gelation process in fabrication CS–TPP nanoparticles in the size range of 100-250 nm intended to be used as carriers for the delivery of gene or protein macromolecules. The critical chitosan concentrations for spontaneous formation of large particle aggregates were found to be 0.65%, 0.25%, 0.15% (w/v) at pH 4.0, and 1.00%, 0.85%, 0.75% (w/v) at pH 5.0 for LMW, MMW and HMW chitosan, respectively. The isoelectric point of the CS–TPP nanoparticles was found at around pH 9.0. Morphological study of the nanoparticles formed under different conditions revealed the formation of regularly shaped polyhydrone particles (pentagon and hexagon), indicating a similar crystallization mechanism during the particle formation and growth process. The bovine serum albumine incorporated particles had a size range of 300-350 nm, doubling the size of CS–TPP particles. The protein incorporated particles possessed a tipical spherical shape and smooth surface.

Interpolymer complexes of chitosan with poly(metacryloylamidobenzoic acid) were described [197]. They were obtained by mixing equimolar amounts of polyelectrolytes with subsequent evaporation of solvents. Further thermal treatment of these complexes at 120°C led to the formation of covalent interchain amide bonds. The complexes had physicochemical properties typical of cross-linked gels, in particular, their degree of swelling correlated with the degree of dehydration of the system in the course of thermal treatment.

The PECs of chitosan and alkylene oxide (ethylene oxide or copolymer of ethylene oxide with propylene oxide) co-polymers with maleic acid were obtained [198-200] and used to obtain pH-and thermally labile films (on-off switches) for selective transport of drugs [198, 199]. These films swell at low pH and temperatures and are capable of reversibly and rapidly changing the degree of swelling in response to the change in the ambient pH and temperature. The introduction of model drugs (salicylic acid, phenol and glucose) into films and their isolation from the latter, depending on the pH (3.8, 6.2, 7.2) and temperature (25 and 50° C), as well as the molecular mass of a drug, were examined. Upon increase in temperatures, decrease in pH from 7.2 to 3.8, and decrease in molecular mass of model substances, the rate of their penetration in films increased, while the rate of release of the drugs increased with pH and temperature. This effect was explained by an increase (with pH) in the electrostatic repulsion between the carboxylic groups of the copolymers of alkylene oxide with maleic acid and the anionic groups of the model compounds (phenol or salicylic acid), and also by an increase in the diffusion coefficient. Moreover, the rate of liberation of active substances depended on the microscopic structure of films and the degree of hydrophobicity of alkylene oxide copolymer samples.

Stoichiometric and nonstoichiometric PECs were prepared with polyethylene glycol-monosuccinate (1:1 mol. ratio) and chitosan [201]. The physicochemical properties of the synthesized PEC's were characterized by using elemental analysis, FTIR, H-1 and C-13 NMR, dissolution behavior, and phase transition phenomenon. Furthermore, some properties of the PECs obtained were analyzed by UV-visible spectrometry, wide-angle X-ray diffraction, DSC, SEM and estimated solubility, and cell viability assay, respectively. The study of MTT assay suggested that the viability of *HepG2 human hepatoblastoma* cells on PEC were increased significantly in accordance with mole ratio of CS.

The conditions of synthesis, as well as the properties and structure of chitosan PECs with polyanionic co-polymers based on maleic acid were studied by Samoilova et al. [202-205]. N-Vinylpyrrolidone, styrene and ethylene were used as comonomers of maleic acid. The complex was also modified by introducing a modified polymer derivative, namely, the copolymer of N-vinylpyrrolidone with monomaleamide containing an amino acid residue of lysine. New approaches were suggested to the preparation of biospecific polymeric materials (affinity sorbents for the isolation of wheat germ lectin [202], thromboresistive coatings for devices contacting blood [205]).

The choice of maleic acid copolymers as co-polyelectrolytes of chitosan for the preparation of PEC was dictated by the regular structure of these polymers (strictly alternating monomer units) and the possibility of transforming maleic acid residues into reactive anhydride groups, which react (under mild conditions) with amino- and hydroxyl-containing ligands, thus providing easy modification of the polyacid before or after complexation. The above properties are not inherent in the chitosan co-polyelectrolytes described in the literature. Moreover, it is certainly of interest to study the formation of these PEC at the molecular (polymer unit) level, taking into account interatomic distances in reactant groups and the conformations of the polymers used [203].

Chapter 5

CONCLUSIONS

This review convincingly demonstrates that chitosan-based PECs are of great practical value and widely used in scientific research, medicine, biotechnology, food and textile industries, etc. The major medical and biotechnological applications of these PECs are summarized in the Table.

Table. Use of chitosan PEC in medicine and biotechnology

Application	Polyanion in PEC	Ref.
Specific (bio)sorbents	Copolymers of maleic acid with N-vinylpyrrolidone, styrene, or ethylene Alginate	202, 204,205 61
Immobilisation of BAC* according to the polyelectrolyte interaction principle	Antigenes β -Lactoglobulin Insulin Tripolyphosphate	206, 207 164 163 196
Films and pervaporation membranes	Poly(4-styrene sulfonate) Sodium alginate Pectin Carboxymethylchitin Polyacrylic acid Carrageenan	73,191 75, 209 78-80 47 48, 49, 168-173 101

Table. (Continued).

Application	Polyanion in PEC	Ref.
Hemocompatible and thromboresistant materials	Copolymer of N-vinylpyrrolidone with N_{α}-lysine monomaleamide	204, 205
	Carboxymethyldextran	112, 210
	Carboxymethylcellulose	88
	Dextran sulfate	110
and prolonged transport of BAC or drugs	Dextran sulfate	120
	Tripolyphosphat	194, 195
	Alginic acid	58, 59, 66, 75, 94, 96-98, 105, 106, 119
	Pectin	95, 99, 82, 99
	Carrageenan	96, 102-104
	Carboxymethylglucomannan	92
	Mucin	165
	Alkylene oxide - maleic acid copolymer	
	Polyvinyl sulfate	198, 199
	Heparin	120
	Poly(ethylene glycol)/heparin	120
	Hualuronic acid	38
	Xanthan	35, 40
	Gum komdagogu	113,116
	Gelatin	93
	Collagen/hyaluronan	155,157
	Poly(acrylic) acid	153
	Poly(γ-glutamic) acid	183,184
	Polyaspartic acid	186
		187
Tissue scaffolds, wound dressing materials, bioconstraction materials for replacement of coverlets, blood vessels, bone tissue, pulp cells regeneration	Glycosaminoglycans	27,30-33,42-46,211-213
	Gelatin	
	Keratin	151, 152,158-160
	Carboxymethylcellulose	162
	Collagen	89
	DNA	161
	Alginate	128
	Alginate/heparin	69, 107
		65
Substrates for cell immobilization or cell adhesion	Alginic acid	63, 72, 215
	Poly (γ–glutamic acid)	215
	Carrageenan Carboxymethylcellulose	215
	Phosphorylated chitosan	215
	Xanthan	52
	DNA	115
	Gelatin	135
	Collagen/gelatin	154
	Polyethyleneglycol-monosuccinate	156
		201

Application	Polyanion in PEC	Ref.
Non-viral vectors of genetic information transfer	DNA	129-131,137-139, 214
Complexation to reduce toxicity	Endotoxin	122-125

* – biologically active compounds.

Chitosan PECs with a wide range of polyanions of natural and synthetic origin have been studied. It was found that complexation between oppositely charged polyelectrolytes is a complicated process, which depends on many parameters. It should be emphasized that chitosan PECs generally have different types of binding, e.g., hydrogen bonding and intermolecular hydrophobic interactions in addition to electrostatic interactions.

The structure and properties of chitosan PECs depend on the nature and structure of polyanions (molecular masses, conformations of macromolecules, charge density and distribution of ionized groups along the polymeric chain), and also on the conditions of PEC formation (pH and solution ionic strength, temperature, concentration, and polymer ratio). Soluble or insoluble and stoichiometric or non-stoichiometric PECs are formed at certain combinations of the above factors.

By varying the type of co-polyelectrolyte and the conditions of the polyelectrolyte reaction, one can obtain chitosan PEC in the form of gels, nano- and microparticles, films, membranes, porous structures and liquid-crystalline dispersions. PECs with natural anionic polyelectrolytes are best studied, but PECs with synthetic polyanions, including block copolymers, are also of great interest because they expand the possibilities for selectively adjusting the properties of the obtained PECs.

REFERENCES

[1] Ravi Kumar, M N V; Muzzarelli, R A A; Muzzarelli, C; Sashiwa, H; Domb, A J. (2004). Chitosan chemistry and pharmaceutical perspectives. *Chem. Rev.,* 104, 6017-6084

[2] Berger, J; Reist, M; Mayer, J M; Felt, O; Gurny, R. (2004). Structure and interaction in chitosan hydrogels formed by complexation on aggregation for biomedical applications, *Eur. J. Pharm. Biopharm.,* 57, 35-52.

[3] Janes, K A; Calvo, P; Alonso, M J. (2001). Polysaccharide colloidal particles as delivery systems for macromolecules. *Adv. Drug Deliv. Rev.,* 47, 83-97.

[4] Il'ina, A. V; Varlamov, V. P. (2005). Chitosan-based polyelectrolyte complexes: a review. *Appl. Biochem. Microbiol.,* 1, 5–11.

[5] Uragami T; Tokura S. Eds. Material science of chitin and chitosan. Japan.Kodansha, *Springer.* 2006.

[6] Jayakumar, R; Prabaharan M. Eds. Current research and developments on chitin and chitosan in biomaterials research. Vol.1. *Research Signpost.* 2008.

[7] Sorlier, P; Denuzière, A; Viton, C; Domard, A. (2001). Relation between the degree of acetylation and the electrostatic properties of chitin and chitosan. *Biomacromolecules,* 2, 765-772.

[8] Zhang, H; Ruhe, J. (2003) Interaction of strong polyelectrolytes with surface-attached polyelectrolyte brushes – polymer brushes as substrates for the layer-by-layer deposition of polyelectrolytes. *Macromolecules,* 36, 6593-6598.

[9] Kabanov, V.A; Yaroslavov, A.A; Sukhishvili, S.A. (1996). Interaction of polyions with cell-mimetic species: physico-chemical and biomedical aspects. *J. Control. Release,* 39, 173-189.

[10] Yaroslavov, A.A; Yaroslavova, E.G; Rakhnyanskaya, A.A; Menger, F.M; Kabanov, V.A. (1999). Modulation of interaction of polycations with negative unilamellar lipid vesicles. *Colloids Surf. B: Biointerfaces,* 16, 29-43.

[11] Chupyatov, A.M; Rogacheva, V.B; Zezin, A.B; Kabanov, V.A. (1994). Kinetics of reactions between oppositely charged linear and network polyelectrolytes. *Polymer.Sci. Ser. A,* 36, 169-174.

[12] Kabanov, V.A; Zezin, A.B; Rogacheva, V.B; Prevish, V.A. (1989). Active transport of linear polyions in oppositely charged swollen polyelectrolyte networks. *Macromol. Chem.,* 190, 2211-2216.

[13] Karpushkin, E.A; Kechek'yan, A.S; Zezin, A.B. (2006). Interpolyelectrolyte reaction between the particles of oppositely charged microgels. *Polymer Science, Ser. B,* 48, 301-303.

[14] Kalsin, A.M; Kowalczyk, B; Smoukov, S.K; Klajn R; Grzybowski, B.A. (2006). Ionic-likc bchavior of oppositely charged nanoparticles. *J. Am. Chem. Soc.,* 128, 15046-15047

[15] Kabanov, V.A. (1994). Physicochemical basis and prospects of using soluble interpolyelectrolyte complexes (review). *Polymer Sci. Ser. A,* 36, 143-156.

[16] Kabanov, V.A; Papisov, I.M. (1979). Formation of complexes between complementary synthetic polymers and oligomers in dilute solution. *Review. Polymer Science USSR. Ser. A,* 21, 261-307.

[17] Zezin, A.B; Kabanov, V.A. (1982). A new class of complex water-soluble polyelectrolytes. *Russ. Chem. Rev.,* 51, 833-855.

[18] Izumrudov, V.A; Zezin, A.B; Kabanov, V.A. (1991). Equlibria in interpolyelectrolyte reactions and the phenomenon of molecular ‘recognition’ in solutions of interpolyelectrolyte complexes. *Russ. Chem. Rev.,* 60, 792-806.

[19] Kabanov, V.A. (2005). Polyelectrolyte complexes in solution and in bulk. *Russ. Chem. Rev.,* 74, 3-20.

[20] Izumrudov, V.A; Sobachin, A.V. (2006). Phase separation in solutions of polyelectrolyte complexes: the decisive effect of a host polyion. *Polymer Science. Ser. A.,* 48, 1098-1104.

[21] Kabanov, V.A; Zezin, A.B. (1984). A new class of complex water-soluble polyelectrolytes. *Makromol. Chem.* (Suppl.), 6, 259-276.

[22] Kabanov, A.V; Vinogradov. S.V; Suzdaltseva,Yu.G; Alakhov, V.Yu. (1995). Water-soluble block polycations as carriers for oligonucleotide delivery. *Bioconjugate Chem.,* 6, 639-643.

[23] Harada, A; Kataoka, K. (1995). Formation of polyion complex micelles in an aqueous milieu from a pair of oppositely-charged block copolymers with poly(ethylene glycol) segments. *Macromolecules,* 28, 5294-5299.

[24] Kabanov, A.V; Bronich, T.K; Kabanov, V.A; Yu, K; Eisenberg, A. (1996). Soluble stoichiometric complexes from poly(N-ethyl-4-vinylpyridinium) cations and poly(ethylene oxide)-block-polymethacrylate anions. *Macromolecules,* 29, 6797-6802.

[25] Kataoka, K; Togawa, H; Harada, A; Yasugi, K; Matsumoto, T; Katayose, S. (1996). Spontaneous formation of polyion complex micelles with narrow distribution from antisense oligonucleotide and cationic block copolymer in physiological saline. *Macromolecules,* 29, 8556-8557.

[26] Kramarenko, E.Yu; Khokhlov, A.R. (2007). Effect of formation of ion pair on the stability of stoichometric block ionomer complexes. *Polymer Science. Ser. A,* 49, 1953-1060.

[27] Denuziere, A; Ferrier, D; Domard, A. (2000). Interactions between chitosan and glycosaminoglycans (chondroitin sulfate and hyaluronic acid): physicochemical and biological studies. *Ann. Pharm. Fr.,* 58, 47-53.

[28] Kuberka, M; Heschel, I; Glasmacher, B; Rau, G. (2002). Preparation of collagen scaffolds and their applications in tissue engineering. *Biomed. Tech.* (Berl), 47, 485-487.

[29] Chatelet, C; Damour, O; Domard, A. (2001). Influence of the degree of acetylation on some biological properties of chitosan films. *Biomaterials,* 22, 261-268.

[30] Shinoda, K; Nakajima, A. (1975). Complex formation of the hyaluronic acid or chondroitin sulfate with glycol chitosan. *Bull. Chem. Res.,* Kyoto Univ., 53, 400-408.

[31] Shinoda, K; Nakajima, A. (1975). Complex formation of the heparin or sulfated cellulose with glycol chitosan. *Bull. Chem. Res.,* Kyoto Univ., 53, 392-399.

[32] Denuziere, A; Ferrier, D; Domard, A. (1996). Chitosan-chondroitin sulfate and chitosan-hyaluronate polyelectrolyte complexes. Physico-chemical aspects. *Carbohydr. Polym.,* 29, 317-323.

[33] Rusu-Balaita, L; Desbrieres, J; Rinaudo, M. (2003). Formation of a biocompatible polyelectrolyte complex: chitosan-hyaluronan complex stability. *Polymer Bull.*, 50, 91-98.

[34] Chen, W.B; Wang, L.F; Chen, J.S; Fan, S.Y. (2005). Characterization of polyelectrolyte complexes between chondroitin sulfate snd chitosan in the solid state. *J. Biomed. Mater. Res.* A, 75, 1238-137.

[35] Catenacci, S.L; Zeccti, V. (2009). Novel mucoadhesive nasal inserts based on chitosan/hyaloranate polyelectrolyte complexes for peptide and protein delivery. *J. Pharmacy and Pharmacology*, 61, 151-157.

[36] Kim, S.J; Shin, S.R; Lee, S.M; Kim, I.Y; Kim, S.I. (2009). Thermal characteristics of polyelectrolyte complexes composed of chitosan and hyaluronic acid. *J. Macromol. Sci.*, 40, 807-815.

[37] Bodohi, S; Moore, N; Johnsonm P.A, Kipper, M.J. (2009). Polysaccharide-based polyelectrolyte complex nanoparticles from chitosan, heparin, and hyaluronan. *Biomacromolecules,* 10, 1402-1409.

[38] Bae, K; Moon, C; Lee, Y.; Park, T. (2009). Intracellular delivery of heparin complexed with chitosan-*g*-poly(ethylene glycol) for inducing apoptosis. *J. Pharmaceutical Research*, 26, 93-100.

[39] Sun, W; Mao, S; Mei, D; Kissel, T. (2008). Self-assembled polyelectrolyte nanocomplexes between chitosan derivatives and enoxaparin. European *J. Pharmaceutics and Biopharmaceutics,* 69, 417-425.

[40] Grech, J.M.R; Mano, J.F; Reis P.L. (2008).Chitosan beads as templates for layer-by-layer assembly and their application in the sustained release of bioactive agents. *J. Bioactive Compatible Polymers,* 23, 367-380.

[41] Hirano, S; Mizutani, C; Yamaguchi, R; Miura, A. (1978). Formation o the polyelectrolyte complexes of some acidic glycosaminoglycans with partially N-acylated chitosans. Communication to the Editor. *Biopolym.*, 17, 805-810.

[42] Denuziere, A; Ferrier, D; Damour, O; Domard, A. (1998). Chitosan-chondroitin sulfate and chitosan-hyaluronate complexes: biological properties. *Biomaterials*, 19, 1275-1285.

[43] Kim, S.J; Yoon, S.G; Lee, K.B; Park, Y.D; Kim, S.I. (2003). Electrical sensitive behavior of a polyelectrolyte complex composed of chitosan/hyaluronic acid. *Solid State Ionics*, 164, 199–204.

[44] Sundararajan, V; Madihally, Howard, W.T. M. (1999). Porous chitosan scaffolds for tissue engineering. *Biomaterials*, 20, 1133 -1142.

[45] Shahabeddin, L; Damour, O; Berthod, F; Rousselle, P; Saintigny, G; Collombel, C. (1991). Reconstructed skin from co-cultured human keratinocytes and fibroblasts on a chitosane cross-linked collagen-GAG matrix. *J. Mater. Sci - Mater. Med.*, 2, 222–226.

[46] Kratz, G; Back, M; Arnander, C; Larm, O. (1998). Immobilised heparin accelerates the healing of human wounds in vivo. *Scand. J. Plast. Reconst. Surg. Hand. Surg.*, 32, 381–385.

[47] Ageev, E.P; Vikhoreva, G.A; Gal'braikh, L.S; Matushkina, N.N; Chaika, E.M; Yaminskii, I.V. (1998). Preparation and properties of films based on chitosan and polyelectrolyte complexes of chitosan with carboxymethylchitin. *Polymer Sci. Ser.* A, 40, 757-764.

[48] Ageev, E.P; Kotova, S.L; Skorikova, E.E; Zezin, A.B. (1996). Pervaporation membranes based on polyelectrolyte complexes of chitosan and poly(acrylic acid). *Polymer Sci. Ser. A,* 38, 202-208.

[49] Skorikova, E.E; Kalyuzhnaya, R.I; Vikhoreva, G.A; Gal'braikh, L.S; Kotova, S.L; Ageev, E.P; Zezin, A.B; Kabanov, V.A. (1996). Properties of interpolyelectrolyte complexes of chitosan and poly(acrylic acid). *Polymer Sci. Ser.* A, 38, 49-53.

[50] Schatz, C; Bionaz, A; Lucas, J.-M; Pichot, C; Viton, C; Domard, A; Delair, T. (2005). Formation of polyelectrolyte complex particles from self-complexation of N-sulfated chitosan. *Biomacromolecules*, 6, 1642-1647.

[51] Bulvan, M; Zapotoczny, S; Nowakowska, M. (2009). Robust one-component chitosan-based ultrathin films fabricated using layer-by-layer technique. *Soft Mater.*, 5, 4726-4732.

[52] Li, Q.-L; Wu, M.-Y; Tang, L.-L; Zhou, J; Jiang, Y; Darvell, B.W. (2008). Bioactivity of a novel nano-composite of hydroxyapatite and chitosan – phosphorylated chitosan polyelectrolyte complex. *J. Bioactive Compatible Polymers,* 23, 520-531.

[53] Mitsumata, T; Suemitsu, Y; Fujii, K; Fujii, T; Taniguchi, T; Koyam K. (2003). pH-response of chitosan, κ-carrageenan, carboxymethyl cellulose sodium salt complex hydrogels. *Polymer,* 44, 7103–7111.

[54] Shumilina, E.V; Shchipunov, Yu.A. (2002). Chitosan-carrageenan gels. *Colloid J.*, 64, 372-378.

[55] Beaumont, M.D; Knorr, D. (1987). Effects of immobilizing agents and procedures on viability of cultured celery (Apiumgraveolens) cells. *Biotechnol. Lett.,* 9, 377-382.

[56] Sakiyama, T; Chu, C.-H; Fujii, T; Yano, T. (1993). Preparation of a polyelectrolyte complex gel from chitosan and κ-carrageenan and its pH-sensitive swelling. *J. Appl. Polym. Sci.,* 50, 2021-2025.

[57] Hugerth, A; Caram-Letham, N; Sundelof, L.O. (1997). The effect of charge density and conformation on the polyelectrolyte complex formation between carrageenan and chitosan. *Carbohydr. Polym.,* 34, 149-156.

[58] Pasparakis, G; Bouropoulas, N. (2006). Swelling studies and in vitro release of verapamil from calcium alginate and calcium alginate – chitosan beads. *Int. J. Pharmaceutics,* 323, 34-42.

[59] Rajedran, A; Basu, S.K. (2009). Alginat-chitosan particulate system for sustained release of nimodipine. *Tropical J. Pharmaceutical Research,* 8, 433-440.

[60] Cekie, N.D; Savic, S.D; Milic, J; Savic, M.M; Jovic, Z; Malesevic, M. (2007). Preparation and characterization of pheniton–loaded alginate and alginate – chitosan microparticles. *Drug Delivery,* 14, 483-490.

[61] Sankalia, M.G; Mashru, R.C; Sankalia, J.M; Satariya, V.B. (2007). Reversed chitosan – alginate polyelectrolyte complex for stability improvement of alfa-amylase: optimization and physicochemical characterization. *European J. Pharmaceutics Biopharmaceutics,* 65, 215-232.

[62] Simsek- Ege, F.A; Bond, G.M; Stringer, J. (2003). Polyelectrolyte complex formation between alginate and chitosan as function of pH. *J. Appl. Polymer Science,* 88, 346-351.

[63] Zhu, H; Xi, Z; Yang, S; Zhang, Y; Wang, H; Guo, H; Zhang, Y; Chen, D; Guo, D.(2009). Responses of free radicals to subcutaneous implantation of alginate – chitosan – alginate (ACA) microcapsules in mice. *International J. artificial organs,* 32, 224-232.

[64] Wattnanaphanit, A; Supaphoi, P; Furuike, T; Tokura, S; Tamura, H; Rujiravanit, R.(2009). Novel chitosan-spotted alginate fibers from wet-spinning of alginate solutions containing emulsified chitosan – citrate complex and their characterization. *Biomacromolecules,* 1, 320-327.

[65] Yim, E.K.F; Liao, I-C; Leong, K.W. (2007). Tissue compatibility of interfacial polyelectrolyte complexation fibrous scaffold evaluation of blood compatibility and biocompatibility. *Tissue Eng.,* 13, 423-433.

[66] Prajapati, B.G; Sawant, K.K. (2009). Poly electrolyte complex of chitosan alginate for local drug delivery. *International J. Chem. Tech. Res.,* 193, 643-648.

[67] Laurie, G; Keen, I; Chandler-Temple, A.F; Rintoul, L; Fredericks, P.M; Drew, B; Grondahl, L. (2007). Interactions between alginate and chitosan biopolymers characterized using FTIR and XPS. *Biomacromolecules,* 8, 2533-2541.

[68] Yan, X.L; Khor, E; Lim, L.Y. (2001). Chitosan – alginate films prepared with chitosan of different molecular weight. *J. Biomed. Mater. Res.,* 58, 358-365.

[69] Wang, L; Khor, E; Wee, A; Lim, L.Y. (2002). Chitosan – alginate PEC membrane as a wound dressing: assessment of incisional wound healing. *J. Biomed. Mater. Res.,* 63, 610-618.

[70] Wandrey, C; Grigorescu, G; Hunkeler, D. (2002). Stydy of polyelectrolyte complex formation applying the synthetic boundary technique of analytical ultracentrifugation. *Progress in Colloid & Polymer Science,* 119, 84-91.

[71] Peng, P; Voelcker, N.H. (2007). Nanoscale eluting coatings based on alginate/chitosan hydrogels. *Biointerphases,* 2, 95-104.

[72] Wu, M.-Y; Chen, N; Liu, L.-K; Yuan, H; Li, Q.-L; Chen, S.-H. (2009). Chitosan/alginate multilayer scaffold encapsulating bone narrow stromal cells in situ on titanium. *J. Bioactive Compatible Polym.*, 24, 301-315.

[73] Berth, G; Voigt, A; Dautzenberg, H; Donath, E; Möhwald, H. (2002). Polyelectrolyte complexes and layer-by-layer capsules from chitosan/chitosan sulfate. *Biomacromolecules,* 3, 579 -590.

[74] Yan, X.L; Khor, E; Lim, L.Y. (2001). Chitosan-alginate films prepared with chitosans of different molecular weights. J. Biomed. Mater. Research (Appl. Biomater.), 58, 358 – 365.

[75] Yan, X.L; Khor, E; Lim, L.Y. Chitosan-alginate PEC films prepared from chitosan of different molecular weights. In: M.G.Peter, A.Domard and R.A.A.Muzzarelli, Eds. *Advan. Chitin Sci.,* Vol.4: University of Potsdam; 2000; 473-478.

[76] Rashidova, S.Sh; Milusheva, R.Yu; Semenova, L.N; Mukhamedjanova, M.Yu; Voropaeva, N.L; Vasilyeva, S; Faizieva, R; Ruban, I.N. (2004). Characteristics of interactions in the pectin–chitosan system. *Chromatographia,* 59, 779-782.

[77] Ofori-Kwakye, K; Fell, J.T; Kipo, S.L. (2006). Effects of pH of medium and molecular weight on polyelectrolyte complex formation between pectin and chitosan. *J. Sci. Technol.* (Ghana), 26, 66-73.

[78] Yao, K.D; Liu, J; Cheng, G.X; Lu, X.D; Tu, H.-L; Silva, J.A.L. (1996). Swelling behavior of pectin/chitosan complex films. *J. Appl. Polym. Sci.,* 60, 279–283.

[79] Bernabé, P; Peniche, C; Argüelles-Monal, W. (2005). Swelling behavior of chitosan/pectin polyelectrolyte complex membranes. Effect of thermal cross-linking. *Polym. Bull.,* 55, 367-375.

[80] Hoagland, P.D; Parris, N. (1996). Chitosan/pectin laminated films. J. Agric. Food Chem., 44, 1915-1919.

[81] Marudova, M (2009). Binding of cationic polyelectrolytes to pectin in solution and in multilayered structures. *J. Optoelectronics and Advanced Materials,* 11, 1428-1431.

[82] Bigucci, F; Luppi, T; Sorrenti, M; Bettinetti, G; Rodrigues, L; Zecchi, V.(2008). Chitosan/pectin polyelectrolyte complexes: selection of suitable preparative conditions for colon-specific delivery of vancomycin. *European J. Pharmaceutical Sci.,* 35, 435-441.

[83] Arguellles-Monal, W; Cabrera, G; Peniche, C; Rinaudo, M. (2000). Conductimetric study of the interpolyelectrolyte reaction between chitosan and polygalacturonic acid. *Polymer,* 41, 2373-2378.

[84] Srinivasan, R; Kamalam, R. (1982). Polyelectrolyte complexes of glycol chitosan with some polysaccharides. 1. Mixing ratio and dielectric properties. *Biopolymers,* 21, 251-263.

[85] Srinivasan, R; Kamalam, R. (1982). Polyelectrolyte complexes of glycol chitosan with some polysaccharides. 2. Electrical conductivite. *Biopolymers*, 21, 265-275.

[86] Arguelles-Monal, W; Garciga, M; Peniche-Covas, C. (1990). Study of the stoichiometric polyelectrolyte complex between chitosan and carboxymethyl cellulose. *Polym. Bull.,* 23, 307–313.

[87] Arguelles-Monal, W; Hechavarria, O.L; Rodrigues, L; Peniche-Covas, C. (1993). Swelling of membranes from the polyelectrolyte complex between chitosan and carboxymethyl cellulose. *Polym. Bull.,* 31, 471–478.

[88] Fukuda, H. (1980). Polyelectrolyte complexes of chitosan with sodium carboxymethylcellulose. *Bull. Chem. Soc. Jpn.,* 53, 837-840.

[89] Chen, H; Fan, M; (2007). Chitosan/carboxymethyl cellulose polyelectrolyte complex scaffolds for pulp cells regeneration. *J. Bioactive Compatible Polymers,* 22, 475-491.

[90] Zhao, L; Mitomo, H; Yosh, F. (2008). Synthesis of pH-sensitive and biodegradable CM-cellulose/chitosan polyampholytic hydrogels with electrone beam irradiation. *J. Bioactive Compatible Polymers,* 23, 319-333.

[91] Maciel, J.S; Silva, D.A; Haroldo, C.B; Paula, R.C.M. (2005). Chitosan/carboxymethyl cashew gum polyelectrolyte complex: synthesis and thermal stability. *Eur. Polym. J.,* 41, 2726–2733.

[92] Du. J; Sun, R; Zhang, S; Zhang, L.F; Xiong, C.D; Peng, Y.X. (2005). Novel polyelectrolyte carboxymethyl konjac glucomannan-chitosan nanoparticles for drug delivery. I. Physicochemical characterization of the carboxymethyl konjac glucomannan-chitosan nanoparticles. *Biopolymers,* 78, 1-8.,

[93] Naidu,V.G.M; Madhusudhana, K; Sashidhar, R.B; Ramakrishna, S; Khar, F.J; Diwan, P.V. (2009). Polyelectrolyte complexes of gum kondagogu and chitosan, as diclofenac carriers. *Carbohydrate Polymers,* 76, 464-471.

[94] Mitrevej, A; Sinchaipanid, N; Rungvejhavuttivittaya, Yu; Kositchaiyong, V. (2001). Multiunit controlled-release diclofenac sodium capsules using complex of chitosan with sodium alginate or pectin. *Pharm. Develop. Technol.,* 6, 385 – 392.

[95] Kim, T.H; Park, Y.H; Kim, K.J; Cho, C.S. (2003). Release of albumin from chitosan-coated pectin beads in vitro. *Int. J. Pharm.***,** 250, 371-383.

[96] Tapia, C; Escobar, Z; Costa, E; Sapag-Hagar, J; Valenzuela, F; Basualto, C; Gai, M.N; Yazdani-Pedram, M. (2004). Comparative studies on polyelectrolyte complexes and mixtures of chitosan–alginate and chitosan–carrageenan as prolonged diltiazem clorhydrate release systems. *Eur. J. Pharm. Biopharm.,* 57, 65–75.

[97] Gonzales-Rodrıguez, M.L; Holgado, M.A; Sanches-Lafuente, C; Rabasco, A.M; Fini, A. (2002). Alginate/chitosan particulate systems for sodium diclofenac release. *Int. J. Pharm.,* 232, 225–234.

[98] Lee, K.Y; Park, W.H; Ha, W.S. (1997). Polyelectrolyte complexes of sodium alginate with chitosan or its derivatives for microcapsules. *J. Appl. Polym. Sci.,* 63, 425–432.

[99] Meshali, M.M; Gab, K.E. (1993). Effect of interpolymer complex formation of chitosan with pectin or acacia on the release behaviour of chlorpromazine HCl. *Int. J. Pharm.,* 89, 177–181.

[100] Muzzarelli, C; Stanic, V; Gobbi, L; Tosi, G; Muzzarelli R A A. (2004). Spray-drying of solutions containing chitosan together with polyuronans and characterisation of the microspheres. *Carbohydr. Polym.,* 57, 73-82.

[101] Park, S.Y; Lee, B.I; Jung, S.T; Park H.J. (2001). Biopolymer composite films based on κ-carrageenan and chitosan. *Mater. Res. Bull.,* 36, 511-519.

[102] Bartkowiak, A; Hunkeler, D. (2001). Carrageenan-oligochitosan microcapsules: optimization of the formation process. *Colloids Surf. B,* 21, 285-299.

[103] Bani-Jaber, A; Al-Ghazawi, M. (2005). Sustained release characteristics of tablets prepared with mixed matrix of sodium carrageenan and chitosan: effect of polymer weight ratio, dissolution medium, and drug type. *Drug Dev. Ind. Pharm.,* 31, 241-247.

[104] Bhise, K.S; Dhumal, R.S; Chauhan, B; Paradkar, A; Kadam, S.S. (2007). Effect of oppositely charged polymer and dissolution medium on swelling, erosion, and drug release from chitosan matrices. *AAPS Pharm. Sci. Tech.,* 8, Article 44.

[105] Mi, F.-L; Sung, H.-W; Shyu, S.-S. (2002). Drug release from chitosan–alginate complex beads reinforced by a naturally occurring cross-linking agent. *Carbohydr. Polym.,* 48, 61–72.

[106] Miyazaki, S; Nakayama, A; Oda, M; Takada, M; Attwood D. (1994). Chitosan and sodium alginate based bioadhesive tablets for intraoral drug delivery. *Biol. Pharm. Bull.,* 17, 745-747.

[107] Kim, H.J; Lee, H.C; Oh, J.S; Shin B.A; Park, R.D; Yang, K.S; Cho, C.S. (1999). Polyelectrolyte complex composed of chitosan and sodium alginate for wound dressing application. *Biomater. Sci. Polym. Ed.,* 10, 543-556.

[108] Sakiyama, T; Takata, H; Kikuchi, M; Nakanishi, K. (1999). Polyelectrolyte complex gel with high pH-sensitivity prepared from dextran sulfate and chitosan. *J. Appl. Polym. Sci.,* 73, 2227-2233.

[109] Kikuchi, Y; Fukuda, H. (1974). Polyelectrolyte complexes of sodium dextran sulfate with chitosan. *Makromol. Chem.,* 175, 3593-3597.

[110] Fukuda, H; Kikuchi, Y. (1977). Polyelectrolyte complexes of sodium dextran sulfate with chitosan. *Makromol. Chem.,* 178, 2895-2899.

[111] Drogoz, A; David, L; Rochas, C; Domard, A; Delair, T. (2007). Polyelectrorolyte complexes from polysaccharides: formation and stoichiometry monitoring. *Langmuir,* 23, 10950-10958.

[112] Kikuchi, Y; Takewayashi, T. (1982). In vitro clot forming properties of polyelectrolyte complexes of [2-(diethylamine) ethyl] dextran hydrochloride with poly(sodium-L-glutamate) and of sodium carboxymethyldextran with chitosan. *Bull. Chem. Soc. Jpn.,* 55, 2307-2308.

[113] Dimitriu, S; Magny, P; Montane, D; Vidal, P.F.(1994).Polyionic hydrogels obtained by complexation between xanthan and chitosan: Their properties as supports for enzyme immobilizations. *J. Bioactive Compatible Polymers*, 9, 184-209.

[114] Chu, C.H; Sakiyama, T; Yano, T. (1995). pH-sensitive swelling polyelectrolyte complex gel was prepared from xantan and chitosan. *Bioscience Biotechnology and Biochemistry,* 59, 717-719.

[115] Chu, C.H; Kumagai, H; Nakamura, K. (1996). Application of polyelectrolyte complex gel composed of xantan and chitosan to the immobilization of *Corynebacterium glutamicum. J. Appl. Polym. Sci.,* 60, 1041-1047.

[116] Dimitriu, S; Chornet, S. (1997). Immobilization of xylanase in chitosan-xanthan hydrogels. *Biotechnology Progress,* 13, 539-545.

[117] Chellat, F; Tabrizian, M; Dumitriu, S; Chornet, S; Rivard, C.H; Yahia, L. (2000). Study of biodegradation behavior of chitosan-xantan microspheres in simulated physiological media. *J. Biomed. Mater. Res.,* 53, 592-599.

[118] Argin Soysal, S; Kofinas, P; Lo, Y.M. (2009). Effect of complexation conditions on xanthan-chitisan polyelectrolyte complex gels. *Food Hydrocolloids,* 23, 202-209.

[119] Kara, F; Demirel, G; Tümtürk, H. (2006). Immobilization of urease by using chitosan-alginate and poly(acrylamide-co-acrylic acid)/kappa-carrageenan supports. *Bioprocess Biosyst. Eng.,* 29, 207-211.

[120] Yamaguchi, R; Arai, J; Hirano, S; Ito, T. (1978). Incorporation of glycoamylase into some polyelectrolyte complexes. *Agric. Biol. Chem.,* 42, 1297-1299.

[121] Brandenburg, K; Andra, J; Muller, M; Koch, M.H.J; Garidel, P. (2003). Physicochemical properties of bacterial glycopolymers in relation to bioactivity. *Carbohydr. Res.,* 338, 2477-2489.

[122] Davidova, V.N; Yermak, I.M; Gorbach, V.I; Solovieva, T.F. (1999). The effect of temperature on the interaction of Yersinia pseudotuberculosis lipopolysaccharide with chitosan. *Membr. Cell. Biol.,* 13, 49-58.

[123] Davydova, V.N; Ermak, I.M; Gorbach, V.I; Krasikova, I.N; Solov'eva T.F. (2000). Interaction of bacterial endotoxins with chitosan.effect of endotoxin structure, chitosan molecular mass, and ionic strength of the solution on the formation of the complex. *Biochemistry* (Moscow), 65, 1082-1090.

[124] Yermak, I.M; Davidova, V.N; Gorbach, V.I; Luk'yanov, P.A; Solov'eva, T.F; Ulmer, A.J; Buwitt-Beckmann, U; Rietschel, E.T; Ovodov Y.S. (2006). Forming and immunological properties of some lipopolysaccharide-chitosan complexses. *Biochimie,* 88, 23-30.

[125] Davydova, V.N; Naberezhnykh, G.A; Ermak, I.M; Gorbach, V.I; Solov'eva T.F. (2006). Determination of binding constants of lipopolysaccharides of different structure with chitosan. *Biochemistry* (Moscow), 71, 332-339.

[126] Torchilin, V.P. (2006). Recent approaches to intracellular delivery of drugs and DNA and organelle targeting. *Ann. Rev. Biomed. Eng*., 8, 343-375.

[127] Yang, F; Cui, X; Yang, X. (2002). Interaction of low-molecular-weight chitosan with mimic membrane studied by electrochemical methods and surface plasmon resonance. *Biophys. Chem.,* 99, 99–106.

[128] Kokushima, T; Hayakawa, T; Okamura, K; Takeda, S; Inouem Y; Miyazaki, K; Okahata, Y. (2006). Buffer solution can control the porosity of chitosan/DNA complexes. *J. Biomed. Mater Res.,* 76B, 121-129.

[129] Ishii, T; Okahata, Y; Sato, T. (2001). Mechanism of cell transfection with plasmid/chitosan complexes. *Biochim. Biophys. Acta,* 1514, 51-64.

[130] Masotti, A; Bordi, F; Ortaggi, G; Marino, F; Palocci, C. (2008). A novel method to obtain chitosan/DNA nanospheres and a study of their release properties. *Nanotechnology.* 19, 055302.

[131] Lavertu, M; Méthot, S; Tran-Khanh, N; Buschmann, M.D. (2006). High efficiency gene transfer using chitosan/DNA nanoparticles with specific combinations of molecular weight and degree of deacetylation. *Biomaterials*, 27, 4815-4824.

[132] Peng, S.-F; Yang, M.-J; Su, C.-J; Chen, H.-L; Lee, P.-W; Wei, M.-C; Sung, H.-W. (2009). Effects of incorporation of poly(γ̃glutamic acid) in chitosan/DNA complex nanoparticles on cellular uptake and transfection efficiency. *Biomaterials,* 30, 1797-1808.

[133] Yuan, Y; Tan, J; Wang, Y, Qian, C; Zhang, M. (2009). Chitosan nanoparticles as non-viral gene delivery vehicles on

atomic force microscopy study. *Acta Biochimica Biophisica Sinica,* 41, Issue11, 515-526.

[134] Xu, S; Dong, M; Liu, X; Howard, K.A; Kjems, J; Besenbacher, F. (2007). Direct force measurements between siRNA and chitosan molecules using force spectroscopy. *Biophys.* J., 93, 952-959.

[135] Kawazoe, N; Narita, Y; Chen, G; Satomi, T; Tateishi, T. (2008). Chitosan/DNA polyelectrolyte complex membranes for controlling cell spreding and aggregation. *The Open Biotechnology J.,* 2, 148-151.

[136] Sato, T; Ishii, T; Okahata, Y. (2001). In vitro gene delivery mediated by chitosan. Effect of pH, serum, and molecular mass of chitosan on the transfection efficiency. *Biomaterials,* 22, 2075-2080.

[137]Liu, W. G; Zhang, X; Sun, S.J; Sun, G. J; Yao, K.D; Liang, D.C; Guo, G; Zhang, J.Y. (2003). N-alkylated chitosan as a potential nonviral vector for gene transfection. *Bioconjug. Chem.,* 14, 782-789.

[138] Wenguang, L; Kangde, Y; Qinggang, L. (2001). Formation of a DNA/N-dodecylated chitosan complex and salts induced gene delivery. *J. Appl. Polym.Sci.,* 82, 3393-3398.

[139] Gao, S; Chen, J; Xu, X; Ding, Z; Yang, Y.-H; Hua, Z; Zhang, J. (2003). Galactosylated low molecular weight chitosan as DNA carrier for hepatocyte-targeting. *Int. J. Pharm.,* 255, 57-68.

[140] Salyanov, V I; Il'ina, A V; Varlamov, V P; Evdokimov, Yu M. (2002). Effect of intercalators on the properties of liquid-crystalline dispersions of nucleic acid – chitosan complex. *Mol. Biol.,* 36, 551-557.

[141] Yevdokimov, Yu.M; Salyanov, V.I. (2003). Liquid crystalline dispersions of complexes formed by chitosan with double-stranded nucleic acids. *Liquid Crystals,* 30, 1057–1074.

[142] Vorob'ev, E.A; Nechipurenko, Yu.D; Salyanov, V.I; Evdokimov, Yu.M. (2007). Description of the binding of chitosan to DNA at different ionic strengths in terms of the theories of ion condensation and adsorption. *Biofizika,* 52, 636-642.

[143] Kruif, K; Weinbreck, F; Vries R. Complex coacervation of proteins and anionic polysaccharides. Lecture notes, NIZO food research BV, Ede Van‘t Hoff Laboratory. The Netherlands: Utrecht University. Wageningen University; 2004.

[144] Teravel, M.N; Domard, A. (1993). Relation between the physico-chemical characteristics of collagen and its interactions with chitosan. Part I. *Biomaterials,* 14, 930-938.

[145] Teravel, M.N; Domard, A. (1995). Collagen and its interaction with chitosan. II. Influence of the physicochemical characteristics of collagen. *Biomaterials,* 16, 865-871.

[146] Chen, J-P; Chang, G-Y; Chen J-K. (2008). Electrospun collagen/chitosan nanofibrous membrane as wound dressing. *Colloids Surf. A.,* 313 – 314, 183-188.

[147] Braudo, E.E; Plashchina, I.G; Schwenke, K.D. (2001). Plant protein interactions with polysaccharides and their influence on legume protein functionality. *A Review. Nahrung/Food,* 45, 382–384.

[148] Plashchina, I.G; Mrachkovskaya, T.A; Danilenko, A.N; Kozhevnikov, G.O; Starodubrovskaya, N.Y; Braudo, E.E; Schwenke, K.D. Complex formation of Faba bean legumin with chitosan: surface activity and emulsion properties of complexes. In: E.Dickinson, R. Miller. Eds. Food colloids, fundamentals of formulation. Cambridge: Royal Society of Chemistry; 2001; 332-341.

[149] Yin, Y; Li, Z; Sun, Y; Yao, K. (2005). A preliminary study on chitosan/gelatin polyelectrolyte complex formation. *J. Materials Sci.,* 40, 4649-4652.

[150] Yao, K; Mao, J; Yin, Y; Liu, W; Cui, Y; Cai, K; Zhao,F. (2002). Chitosan/gelatin network based biomaterials in tissue engineering. *Biomed. Engineering Applications, Basis&Communications,* 14, 115-121.

[151] Shen, F; Cui, Y.L; Yang, K.D; Yao, X.H. (2000). A study on the fabrication of porous chitosan/gelatin network scaffold for tissue engineering. *Polymer Intern.,* 49, 1596-1599.

[152] Zhao, F; Yin, Y; Lu, W.W; Leong, J.C; Zhang, J; Zhang, M; Yao, K. (2002). Preparation and histological evaluation of biomimetic three-dimensional hydroxyapatite/chitosan-gelatin network composite scaffold. *Biomaterials*, 23, 3227-3234.

[153] Lin, Y.-C; Tan, F.-J; Marra, K.G; Jan, S.-S; Liu, D.-C. (2009). Synthesis and characterization of collagen/hyaluronan/chitosan composite sponges for potential biomedical applications. *Acta Biomater.,* 5, 2591-2600.

[154] Zhu, A; Zhao, F; Ma, T. (2009). Photo-initiated grafting of gelatin/N-maleic acyl-chitosan to enhance endothelial cells adhesion, proliferation and function on PLA surface. *Acta Biomater.,* 5, 2033-2044.

[155]Huang, Y; Onyeri, S; Siewe, M; Moshfeghian, A; Modihally, S.V. (2005). In vitro characterization of chitosan-gelatin scaffolds for tissue engineering. *Biomaterials,* 26, 7616-7627.

[156]Yu, B.-Y; Chou, P.-H; Chen, C.-A; Sun, Y.-M; Kang, S.-S. (2007). C3A cell behaviors on micropatterned chitosan-collagen-gelatin membranes. *J. Biomaterials Applications,* 22, 255-274.

[157]Yao, K.D; Yin, Y.J; Xu, M. X; Wang, Y.F. (1995). Investigation of pH-sensitive drug delivery system of chitosan/gelatin hybrid polymer network. *Polym. Intern.,* 38, 77-82.

[158] Shen, F; Cui, Y.L; Yang, L.F; Yao, K.D; Dong, X.H; Jia, W.Y; Shi, H.D. (2000). A study on the fabrication of porous chitosan/gelatin network scaffold for tissue engineering. *Polym. Int.,* 49, 1596 – 1599.

[159] Mao, J.S; Zhao, L.G; Yin, Y.J; Yao, K.D. (2003). Structure and properties of bilayer chitosan – gelatin scaffolds. *Biomaterials,* 24, 1067–1074.

[160] Zhang, L; Ao, Q; Wang, A; Lu, G; Kong, L; Gong, Y; Zhao, N; Zhang, X.A. (2006). Sandwich tubular scaffold derived from chitosan for blood vessel tissue engineering. *J. Biomed. Mater. Res. A*., 77, 277-84.

[161] Zhu, C; Fan, D; Ma, X; Xue, W; Yu, Y; Luo, Y; Liu,B; Chen,L. (2009). Effects of chitosan on properties of novel human-like collagen/chitosan hybrid vascular scaffold. *J. Bioactive Compatible Polym.,* 24, 560-576.

[162] Tanabe, T; Okitsu, N; Tachibana, A; Yamauchi K. (2002). Preparation and characterization of keratin – chitosan composite film. *Biomaterials,* 23, 817-825.

[163]Mao, S; Bakowsky, U; Jintapattanakit, A; Kissel, T. (2006). Self-assembled polyelectrolyte nanocomplexes between chitosan derivatives and insulin. *J. Pharmaceutical. Sci*., 95, 1035-1048.

[164] Casal, E; Montilla, A; Moreno, F.J; Olano, A; Corzo, N. (2006). Use of chitosan for selective removal of □lactoglobulin from whey.*J Dairy Sci.,* 89, 1384-1389.

[165] Beeley, J.G. Glycoprotein and Roteoglycan Techniques. In: R.H.Burdon, P.H.van Knippenberg. Eds. (Laboratory techniques in biochemistry and molecular biology; v.16) Amsterdam - New-York - Oxford: Elsevier; 1985; 16-18.

[166] Qaqish, R.B; Amiji, M.M. (1999). Synthesis of a fluorescent chitosan derivative and its application for the study of chitosan-mucin interaction.

[167] Perez-Gramatges, A; Argelles-Monal, W; Peniche-Covas, C. (1996). Thermodynamics of complex formation of polyacrylic acid with poly(N-vinyl-2-pyrrolidone) and chitosan. *Polym. Bull.,* 37, 127-134.
[168] Chavasit, V; Kienzle-Sterzer, C; Torres J.A. (1988). Formation and characterization of an insoluble polyelectrolyte complex: chitosan-polyacrylic acid. *Polym. Bull.,* 19, 223-230.
[169] Skorikova, Ye; Vikhoreva, G.A; Kalyuzhnaya, R.I.; Zezin, A.B; Gal'braikh, L.S; Kabanov, V.A. (1988). Polyelectrolyte complexes based on chitosan. *Polym. Sci. USSR,* 30, 49–55.
[170] Wang, H; Li, W; Lu, Yu; Wang, Zh; Zhong, W. (1996). Studies on chitosan and poly(acrylic acid) interpolymer complex. II. Solution behaviors of the mixture of water-soluble chitosan and poly(acrylic acid*). J. Appl. Polym. Sci.,* 61, 2221-2224.
[171] Shieh, J.J; Huang, R.U.M. (1987). Pervaporation with chitosan membranes. II. Blend membranes of chitosan and polyacrylic acid and comparison of homogeneous and composite membrane based on polyelectrolyte complexses of chitosan and polyacrylic acid for the separationof ethanol-water mixtures. *J. Membr. Sci.,* 127, 185-202.
[172] Nam, S.Y; Lee, V.M. (1997). Pervaporation and properties of chitosan-poly(acrylic acid) complex membranes. *J. Membr. Sci.,* 135, 161-171.
[173] Wang, H; Li, W; Lu, Yu; Wang, Zh. (1997). Studies on chitosan and poly(acrylic acid) interpolymer complex. I. Preparation, structure, pH-sensitivity, and salt sensitivity of complex-forming poly(acrylic acid): chitosan semi-interpenetrating polymer network. *J. Appl. Polym. Sci.,* 65, 1445-1450.
[174] Crescenzi, V; Paradossi, G; Desideri, P; Dentini, M; Cavalieri, F; Honici, E; Lisi, R. (1997). New hydrogels based on carbohydrate and on carbohydrate-synthetic polymer networks. *Polym. Gels Networks,* 5, 225-239.
[175] Staikos, G; Tsitsilians, C. (1991). Viscometric investigation of the poly(acrylic acid)-polyacrylamide interpolymer association. *J. Appl. Polym. Sci., 42*, 867-872.
[176] Hoffmann, H; Kastner, U; Donges, R; Ehrel, R. (1996). Gels from modified hydroxyethyl cellulose and ionic surfactants. *Polym. Gels Networks,* 4, 509-526.
[177] Kobayashi, K; Tsuchida, A; Usui, T; Akaike, T. (1997). A new type of artificial glycoconjugate polymer: a convenient synthesis and its interaction with lectins. *Macromolecules,* 30, 2016-2020.

[178] Cerrai, P; Guerra, G.D; Tricoli, M; Maltinti, S; Barbani, N; Petarca, L. (1996). Polyelectrolyte complexes obtained by radical polymerization in the presence of chitosan. *Macromol. Chem. Phys.*, 197, 3567-3579.

[179] Peniche, C; Arguelles-Monal, W; Davidenko, N; Sastre, R; Gallardo, A; San Roman, J. (1999). Self-curing membranes of chitosan/PAA IPNs obtained by radical polymerization: preparation, characterization, and interpolymer complexsation. *Biomaterials,* 20, 1869-1878.

[180] Torrado, S; Prada, P; de la Torre, P.M; Torrado, S. (2004). Chitosan – poly(acrylic) acid polyionic complex: in vivo study to demonstrate prolonged gastric retention. *Biomaterials,* 25, 917-923.

[181] Wang, J.-W; Chen, C.-Y; Kuo, Y.-M. (2009). Chitosan – poly(acrylic acid) nanofiber networks prepared by the doping induction of succinic acid and its ammonia-response studies. *Polymers for Advanced Technologies,* 19, 1343-1352.

[182] Smitha, B; Sridhar, S; Khan, A.A. (2004). Polyelectrolyte complexes of chitosan and poly(acrylic acid) as proton exchande membranes for fuel cells. *Macromolecules,* 37, 2233-2239.

[183] Oliveira De H.C.L; Fonseca, J.L.C, Pereira, M.R. (2008). Chitosan – poly(acrylic acid) polyelectrolyte complex membranes: preparation, characterization and permeability studies. *J. Biomaterials,* 19, 143-160.

[184] Yadav, KSH; Satish, CS; Shivakumar, HG. (2007). Preparation and evaluation of chitosan – poly(acrylic acid) hydrogels as stomach specific delivery for amoxicillin and metronidazole. *Indian J. Pharmaceutical Sci.,* 69

[185] Zhang, J; Yuan, K; Wang, Y.-P; Zhang, S.-T; Zhang, J. (2007). Preparation and pH responsive behavior of poly(vinyl alcohol) – chitosan – poly(acrylic acid) Full-IPN hydrogels. *J. Bioactive Compatible Polymers,* 22, 207-218.

[186] Kang, H.S; Park, S.H; Lee, Y.G; Son, T.I. (2007). Polyelectrolyte complex hydrogel composed of chitosan and poly(□-glutamic) acid for biological application: Preparation, physical properties, and cytocompatibility. *J. Appl. Polym. Sci.,* 103, 386-394.

[187] Zheng, Y; Yang, W; Wang, C; Hu, J; Fu, S; Dong, L; Wu, L; Shen, X. (2007). Nanoparticles based on the complex of chitosan and polyaspartic acid sodium salt: Preparation, characterization and the use for 5-fluorouracyl delivery. *European J. Pharmaceutics Biopharmaceutics,* 67, 621-631.

[188] Ehtezazi, T; Govender, T; Stolnic, S. (2000). Hydrogen bonding and electrostatic interaction of as cationic drug with polyaspartic acid, *Pharm. Res.,* 17, 871-878.

[189] Jiang, H.L; Zhu, K.J. (2006). Comparison of poly(aspartic acid) hydrogel and poly(aspartic acid)/gelatin complex for entrapment and pH-sensitive release of protein drugs. *J. Appl. Polym. Sci.,* 9, 2320-2329.

[190] Hattori, T; Katai, K. (2008). pH-potentiometric study of polyion complexes between chitosan and poly(vinyl sulfate) or dodecylbenzene sulfonate. *International J. of the Japan Soc. Analytical Chem.,* 24, 543-546.

[191] Budd, P.M; Ricardo, N; Jafa, J.J; Stephenson, B; Hughes, R. (2004). Zeolite/polyelectrolyte multilayer pervaporation membranes for enhanced reaction yield. *Ind. Eng. Chem. Res.,* 43, 1863-1867.

[192] Volkova, I.F; Gorshkova, M.Yu; Izumrudov, V.A. (2008). Water-soluble nonstoichiometric polyelectrolyte complex chitosan with a polystyrene sulfonate anion. *Polymer Science. Ser.* A., 50, 971-976.

[193] Stoilova, O; Koseva, N; Petrova, Ts; Manolova, N; Rashkov, I; Naydenov, M. (2001). Hydrolysis of chitosan, chitosan-polyoxyethylene and chitosan - poly(2-acrylamido-2-methylpropansulfonic acid by a crude enzyme complex from Trichoderma viride. *J. Bioactive Compatible Polym.,* 16, 379-392.

[194] Betigeri, S.S; Neau, S.H. (2002). Immobilization of lipase using hydrophilic polymers in the form of hydrogel beads. *Biomaterials,* 23, 3627–3636.

[195] Alsarraa, I.A; Betigeria S.S; Zhanga, H; Evansb, B.A; Neau, S.H. (2002). Molecular weight and degree of deacetylation effects on lipase-loaded chitosan bead characteristics. *Biomaterials,* 23, 3637–3644.

[196] Gan, Q; Wang, T; Cochrane, C; McCarron, P. (2005). Modulation of surface charge, particle size and morphological properties of chitosan-TPP nanoparticles intended for gene delivery. *Colloids and Surfaces B: Biointerfaces,* 44, 65-73.

[197] Peniche, C; Elvira, C; San Roman, J. (1998). Interpolymer complexes of chitosan and polymethacrylic derivatives of salicylic acid: preparation, characterization and modification by thermal treatment. *Polymer,* 39, 6549-6554.

[198] Yoshizawa, T; Shin-ya, Y; Hong, K.J; Kajiuchi, T. (2004). pH- and temperature-sensitive permeation through polyelectrolyte complex films

composed of chitosan and polyalkyleneoxide–maleic acid copolymer. *J. Membr. Sci.,* 241, 347-354.

[199] Yoshizawa, T; Shin-ya, Y; Hong, K-J; Kajiuchi T. (2005). pH- and temperature – sensitive realease behaviors from polyelectrolyte complex films composed of chitosan and PAOMA copolymer. *Eur. J. Pharm. Biopharm.,* 59, 307-313.

[200] Yoshizawa, T; Shinya, Y; Hong, K; Kajiuchi, T. (2004). Preparation and stimuli-sensitive characterization of polyelectrolyte complex films derived from chitosan and polyalkyleneoxide-maleic acid copolymer. *Trans. Mater. Res. Soc. Jpn.,* 29, 2631-2634.

[201] Son, T; Kim, D; Kim, B; Koo, K; Park, Y; Kim, J; Lim, H. (2007). Stoichiometric and nonstoichiometric polyelectrolyte complex of chitosan and polyethyleneglycol-monosuccinate: preparation and characterization. *J. Appl. Polym. Sci.,* 104, 3057-3070.

[202] Samoilova, N.A; Krayukhina, M.A; Yamskov I.A. (2002). Wheat Germ Lectin Sorption by Chitosan Polyelectrolyte Complexes. *Appl. Biochem. Microbiol.,* 38, 447-451.

[203] Samoilova, N.A; Krayukhina, M.A; Yamskov, I.A. (2002). New ion-exchange and affine sorbents based on polyelectrolyte chitosan complexes. *Russian J. Phys. Chem.,* 76, 1502-1507.

[204] Samojlova, N.A., Krayukhina, M.A.; Yamskov I.A. (2004). Use of the affinity chromatography principle in creating new thromboresistant materials. *J. Chromatography B, Anl. Tech. in the Biomed. & Life Sci.,* 800, 263-269.

[205] Samoilova, N.A; Krayukhina, M.A; Novikova, S.P; Babushkina, T.A; Volkov, I.O; Komarova, L.I; Moukhametova, L.I; Aisina, R.B; Obraztsova, E.A; Yaminsky, I.V; Yamskov, I.A. (2007). Polyelectrolyte thromboresistant affinity coatings for modification of devices contacting blood. *J. Biomed. Mater. Res.,* 82A, 589-598.

[206] Vila, A; Svnchez, A; Janes, K; Behrens, I; Kissel, T; Vila Jato, J.L; Alonso M.J. (2004). Low molecular weight chitosan nanoparticles as new carriers for nasal vaccine delivery in mice. *Eur. J. Pharm. Biopharm.,* 57, 123-131.

[207] Illum, L; Jabbal-Gill, I; Hinchcliffe, M; Fisher, A.N; Davis S.S. (2001). Chitosan as a novel nasal delivery system for vaccines. *Adv. Drug Deliv. Rev.,* 51, 81-96.

[208] Wang, S; Ye, S; Dai, L; Liu, X; Tong Z. (2007). Enhanced resistance of polyelectrolyte multilayer microcapsules to pepsin erosion and release

properties of encapsulated indomethacin. *Biomacromolecules,* 8, 1739 - 1744.

[209] Kim, S-G; Lee, K-S; Lee, K-H. (2006) Pervaporation separation of sodium alginate/chitosan polyelectrolyte complex composite membranes for the separation of water/alcohol mixtures: Characterization of the permeation behavior with molecular modeling techniques. *J. Appl. Polym. Sci.,* 103, 2634–2641.

[210] Fukuda, H; Kikuchi, Y. (1978). Polyelectrolyte complexes of chitosan with sodium carboxymethyldextran. *Bull. Chem. Soc. Jpn.,* 51, 1142-1144.

[211] Suh, J-K.F; Matthew, H.W.T. (2000). Application of chitosan-based polysaccharide biomaterials in cartilage tissue engineering: a review. *Biomaterials,* 21, 2589-2598.

[212] Gutowska, A; Jeong, B M; Jasionowski, M. (2001). Injectable gels for tissue engineering. *Anat. Rec.,* 263, 342-349.

[213] Majima, T; Irie, T; Sawaguchi, N; Funakoshi, T; Iwasaki, N; Harada, K; Minami, A; Nishimura S-I. (2007). Chitosan-based hyaluronan hybrid polymer fibre scaffold for ligament and tendon tissue engineering. *J. Eng. Med.,* 221, 537-546.

[214] Maurstad, G; Danielsen, S; Stokke B T. (2007). The influence of charge density of chitosan in the compaction of the polyanions DNA and xanthan. *Biomacromolecules,* 8, 1124-1130.

[215] Mansfeld, J; Dautzenberg. Immobilization of cells in polyelectrolyte complexes. In: Bickerstaff, G.F. Editor. *Immobilization of enzymes and cells.* Totowa. New Jersy: Humana Press; 1997; 309-317.

INDEX

A

B

C

D

H

I

J

K

L

M

N

O

P

R